Patrick Moore's
Practical Astronomy Series

Other Titles in this Series

Practical Amateur Spectroscopy

Stephen F. Tonkin (Ed.)

With 125 Figures
including 3 in Colour

Springer

Cover illustrations: Meade LX200 courtesy of Meade Instruments Corporation. Stellar Spectral Sequence courtesy of the National Optical Astronomy Observatory/Association of Universities for Research in Astronomy/National Science Foundation (www.noao.edu/image-gallery). Reflection Nebulosity (NGC 1435 and IC 349) around Merope (23 Tau) in the Pleiades (M45), © Yuugi Kitahara.

British Library Cataloguing in Publication Data
Practical amateur spectroscopy. – (Patrick Moore's practical astronomy series)
   1. Astronomical spectroscopy – Amateurs' manuals
   I. Tonkin, Stephen F., 1950–
   522.6'7
ISBN 1852334894

Library of Congress Cataloging-in-Publication Data
A catalog record for this book is available from the Library of Congress

Patrick Moore's Practical Astronomy Series ISSN 1617-7185
ISBN 1-85233-489-4

Printed on acid-free paper

Printed in Singapore   (KYO)

9 8 7 5 6 4 3

Springer Science+Business Media, LLC

springer.com

Acknowledgements

Thanks are due to John Watson of Springer for his ongoing encouragement and support; Mike Nugent who copy-edited the manuscripts; Nick Wilson (also of Springer) for, yet again, his patient help and advice during the production stage; the Santa Barbara Instrument Group of Santa Barbara, CA, and Oriel Instruments Corporation of Stratford, CT, for permission to use material from their catalogues; Valerie Desnoux and Martin Peston for permission to reproduce material related to the use of *Visual Spec* in Chapter 6; and Neil Martin for the spectral profiles in Chapters 4 and 6.

Additionally, we would like to recognise the contribution of those family members, friends and colleagues without whose suggestions, encouragement, and endless cups of coffee, this book would not have been possible.

Photographs and Drawings

All photographs, images and drawings are by the authors unless otherwise credited. It is inevitable that some drawings will have been influenced by those in earlier works on wave optics and spectroscopy; this has been acknowledged where we are conscious of that influence.

Contents

Introduction

Spectroscopy is possibly the most potent technique available to the astronomer. but is probably the one whose possibilities has been least explored by the amateur. It can give us information about the age, composition, distance, expansion or contraction, ionization level, line-of-sight velocity, luminosity, mass, rotational velocity, and temperature of an astronomical object.

Largely because of the very long exposures required to record a spectrum on photographic film, it is a discipline that has, with a few notable exceptions, been ignored by the bulk of the amateur astronomical community. Recently the increased sensitivity of CCD cameras with respect to photographic emulsion makes spectroscopy possible with telescopes of 5 inches aperture and above. Because the CCD camera makes a digital record of the spectrum, it is easily amenable to processing and analysis on a home computer. There are now several powerful application programs, including freeware ones, that facilitate spectral analysis.

The underlying physics is often thought to lie in the realms of the arcane but, as the first section of this book shows, it is easily accessible to anyone with a fundamental understanding of high school physics and mathematics who is prepared to spend a few hours on a cloudy night swotting up the basics. The fundamentals in this part of the book are intended as a basic outline, sufficient to serve as a basis for understanding the underlying principles, and as a springboard to further reading for those who wish to pursue any of these aspects at a deeper level. For a wealth of detail on spectral types, anyone interested in pursuing this avocation should at the very least obtain James Kaler's superb *Stars and Their Spectra*.[1]

Commercial astronomical spectrographs are often considered to be expensive. The more sophisticated

---

[1] Publication details in the bibliography.

ones certainly are when compared to many of the other accessories that an amateur astronomer may wish to use. However, when we consider the increased amount of scientific work that they make available to us, as Dale Mais demonstrates in Chapter 9, they are *pro rata* not as pricey as they first appear. They are also not the only option available to us.

Anything that disperses light has some potential as a spectroscope. My first experiments were with vinyl EP records as crude diffraction gratings. The modern equivalent of this, as David Randell shows in Chapter 5, is the use of a CD or DVD as a grating. This permits some of the grosser spectral characteristics of bright objects to be studied and adequately serves as a zero-cost introduction to basic spectroscopy.

The next step-up is a simple direct vision star spectroscope, which costs no more than a decent eyepiece. Jack Martin has refined his technique with a star spectroscope coupled to his undriven Dobsonian telescope, using star drift to widen the spectrum and make the lines visible and a simple inexpensive SLR camera to record these spectra. His superb results with simple equipment should be an inspiration to any budding spectrographer.

For those wishing to progress beyond what is possible with a star spectroscope, the options are either to use, as Steve Dearden does with great success, instruments designed for other purposes or to make your own. Both Nick Glumac and Tom Kaye have followed this latter route and their results show the extremely high quality of scientific work that can be achieved.

I hope that this book will serve as an introduction to a fascinating aspect of amateur astronomy that recent advances in technology and computing have put within the grasp of all of us. I have emphasized the variety of practical options that are available in the hope that, at the very least, it will inspire you to emulate the excellent work of at least one of the contributors.

Stephen Tonkin
Calgary, January 2002

Section 1

Fundamentals of Spectroscopy

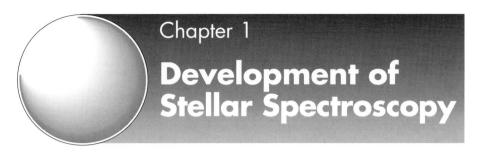

Chapter 1

Development of Stellar Spectroscopy

Stephen Tonkin

The father of stellar spectroscopy is Josef von Fraunhofer (1787–1826), the inventor of the German equatorial mount that is so popular with modern amateur astronomers. He was an optician and, in order to measure the dispersive power of lenses, he made his first spectrometer when he was in his mid-twenties. In 1814, he found that there were lines in the spectrum of white light and, out of interest, he turned his instrument to the Sun. Over the next few years, he mapped the positions of 574 dark lines in the solar spectrum,[1] and labeled the most prominent of these with letters, beginning the alphabet at the red end of the spectrum. This alphabetic classification survives today when, for example, we refer to the "calcium H line" (3968 Å). He found that Venus' spectrum has the same dark lines as the solar spectrum, offering further evidence (if it were needed!) that Venus shines with reflected sunlight. Fraunhofer also turned his objective-prism spectroscope to several of the brighter stars and observed clear differences amongst them. Fraunhofer's other great contribution to modern amateur spectro-scopy was his invention of the diffraction grating.

Fraunhofer was not able to ascribe a cause to these dark lines in the various spectra that he studied, so his contribution in this realm was confined to the

---

[1] William Hyde Wollaston (1766–1828) first recorded the existence of seven of these dark lines over a decade before Fraunhofer, but he did not recognize their significance, attributing them instead to "natural boundaries between the colours". Fraunhofer was unaware of Wollaston's work.

important one of accumulation of data. The explanation of these dark lines was published in 1859 by Gustav Kirchhoff (1824–1887), who attributed the cause to the selective absorption of light by elements in the Sun's atmosphere. In collaboration with Robert Bunsen (1811–1899), whose eponymous burner[2] permitted the study of the spectra of different substances, Kirchhoff continued his work on spectroscopy and encapsulated its basic principles into his three laws:

1. An incandescent solid or a gas under high pressure will produce a continuous spectrum.
2. An incandescent gas under low pressure will produce an emission-line spectrum.
3. A continuous spectrum viewed through a low-density gas at low temperature will produce an absorption-line spectrum.

The technique of spectroscopy soon led to the discovery of new elements (e.g. rubidium, whose name means "red") and in 1868 Norman Lockyer (1836–1920) attributed a line in the solar spectrum to a hitherto unknown element that he named for the Sun: helium. Nearly three decades later helium was isolated on Earth by the Scots chemist William Ramsay. The value of spectroscopy as an investigative astronomical technique was thus further strengthened.

Lockyer's discovery did lead to some false trails. William Huggins (1824–1910) found some emission lines in the Great Orion Nebula that he attributed to an element, "nebulium", that was yet to be discovered on Earth. The lines were later found by Bowen to come from doubly ionized oxygen [O III]. Similarly, lines attributed to a new element, "coronium", in the solar corona were shown to be due to metals in a very high state of ionization.

Huggins was, however, together with the Italian astronomer-priest, Father Pietro Angelo Secchi (1818–1878), among the first to take a systematic approach to stellar spectroscopy. Secchi developed a system of spectral classification that was both simple and powerful:

Type I   –   Simple spectra from blue-white stars like Vega, whose lines are due to hydrogen and helium (although Secchi did not know this).

---

[2]The Bunsen burner is actually an improvement, made not by Bunsen but by one of his assistants whose name is lost to us, of a burner invented by Michael Faraday (1791–1867).

Type II – Multilined spectra like that of the Sun. The multiple lines are now known to be due to metals.

Type III – Spectra with "flutings" (bands of lines) that darken towards blue, found in orange-red stars such as Betelgeuse. These bands have been traced to TiO.

Type IV – Spectra with "flutings" that darken towards red, which Secchi himself attributed to carbon compounds.

Type V – Spectra with bright (emission) lines.

As spectroscopes improved and more detail became apparent in stellar spectra, it became clear that Secchi's classification types needed to be subdivided. The then director of the Harvard College Observatory, Edward Pickering (1846–1919), undertook this work. Pickering's work was financed by the estate of the late Henry Draper and the resulting catalogue was named the *Henry Draper Catalogue* in his honor; stars are still often referred to by their HD (Henry Draper) numbers. The original system of classification devised by Pickering in 1890 and used in the *Draper Catalogue of Stellar Spectra* published in that year was:

A – White or bluish-white stars with dominant hydrogen lines

B – White or bluish-white stars with dominant hydrogen and helium lines

C – White or bluish-white stars with doubled hydrogen lines

D – White or bluish-white stars with emission lines present

E – Ca II (Fraunhofer H and K) and hydrogen lines

F – Like E, but with more hydrogen lines

G – Yellow stars, like F but with many additional metals present

H – Like G, but with a drop in intensity in the blue part of the spectrum

I – Like H but with additional lines

K – Orange stars with visible bands (metals) in spectrum

L – Peculiar variations of K

M – Orange-red stars with spectra complicated by the presence of molecules

N – Red stars with C dominant

O – Mainly bright lines (Wolf–Rayet)

P – Planetary nebulae
Q – All other spectra
(Q was changed to designate novae in 1922.)

Although Pickering was responsible for the work, the actual classification was performed by three women: Williamina Fleming (1857–1911) who was responsible for most of the classifications in the 1890 catalogue, Antonia Maury (1866–1952) and Annie Cannon (1863–1941). Pickering and Fleming refined and simplified the classification, removing some of the classification letters in the process. Maury devised a system that was important in that it not only classified lines with lower-case letters according to their characteristics, but also suggested that it would be logical for O and B to precede A in her 22-category classification system. Maury's system was short-lived, but some of its ideas lived on in the work of Cannon and Ejnar Hertzsprung (1873–1967).

In 1901, Annie Cannon greatly simplified the classification, reducing it to the now-familiar seven letters, OBAFGKM, and adding precision by a decimal classification that plotted the positions of stars between two defined letters. For example, a star whose characteristics lay between those of B and A would be a B5 star (originally B5A, but the last letter was redundant and therefore dropped). She also introduced a lower-case letter classification, different from Maury's system, for classifying the characteristics of the bright lines in type O stars.

Cannon's classification system lives on in the familiar Hertzsprung–Russell (H-R) diagram (Fig 4.1 in Chapter 4), a two-dimensional classification of stars according to their spectral type and their absolute magnitudes (luminosities) developed independently by Ejnar Hertzsprung in 1906 and Henry Norris Russell (1877–1957) in 1910. The position of a star on the H-R diagram depends upon two factors: its mass and its age. The distribution of stars in the diagram is therefore useful in studies of stellar evolution. The characteristics of each spectral class are detailed in Chapter 4.

The two-dimensional classification of the H-R diagram is useful, but incomplete. It is usually applicable to stars in the Galactic disk with similar chemical compositions to that of the Sun (population I stars), but is clearly less so for the older population II stars of the Galactic halo, which have a lower abundance of heavy elements. The great variety of chemical composition has made it difficult to develop a straightforward and consistent system for adding this third dimension to the H-R diagram.

In addition to its use in classification of stars, spectral analysis can also be used to glean information about the motion of astronomical objects. The shift of lines in a spectrum is an indication, caused by the Doppler effect, of whether the object is approaching or receding from the observer. Lines shifted to the blue/ violet end of the spectrum (blue shift) indicate approach; lines shifted to the red end of the spectrum (red shift) indicate recession. Stars were found to have radial velocities[3] of up to about 20 km s^{-1}. By 1912 the photographic and spectrographic tools available were sufficient to allow Vesto Slipher (1875–1969) to analyze the spectra of nebulae, sometimes with exposures of over 36 hours per photographic plate! His first target was the great "spiral nebula" in the constellation of Andromeda (M31, the Great Andromeda Galaxy). The spectrum showed lines characteristic of starlight, but shifted towards the blue/violet end of the spectrum. The degree of blue shift indicated a radial velocity of some 300 km s^{-1}, the greatest known for any object, and evidence that the nebula was outside our Galaxy. Within another four years Slipher had found the radial velocities of 25 "spiral nebulae"; some of these velocities were in excess of 1000 km s^{-1}. It was this pioneering work of Slipher's that laid the foundation for Edwin Hubble's (1889–1953) discovery that the universe is expanding and his proof that many objects classified as "nebulae" are other galaxies. Within its first century of existence, the new science of stellar spectroscopy had more than proved its value.

Bibliography and References

Hoskin, M (1999) *The Cambridge Concise History of Astronomy*. Cambridge University Press, ISBN 0521576008.
Ridpath, I (1997) *A Dictionary of Astronomy*. Oxford University Press, ISBN 0192115960.
Sullivan, N (1965) *Pioneer Astronomers*. Scholastic Book Services, New York.

---

[3]Radial velocity is the component of a velocity along the observer's line of sight. When measuring the radial velocities of objects, the Earth's orbital velocity must be taken into account.

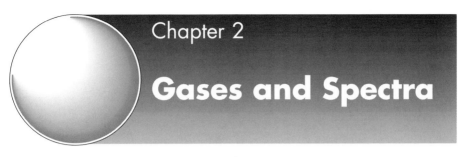

Chapter 2

Gases and Spectra

Stephen Tonkin

The Structure of the Atom

The model of the atom that is useful in spectroscopy is that in which it is subdivided into three principal subatomic particles: *proton, neutron* and *electron*. The multiplicity of other subatomic particles that have been discovered by physicists are important for the under-standing of cosmology and of atomic physics, but their consideration is not necessary for the understanding of the formation of spectra. The two properties of the three principal particles that we need to consider are their mass and their electric charge, as shown in Table 2.1.

The atom consists of a massive positively charged nucleus of neutrons and electrons surrounded by a cloud of negatively charged electrons (Fig. 2.1). It is convenient to consider that the electrons are in orbital shells that surround the nucleus, and that each of these shells is itself subdivided into "subshells" or "suborbitals".

Table 2.1. Mass and charge of subatomic particles

	Mass (kg)	Charge (C)
Proton	1.672×10^{-27}	1.602×10^{-19}
Neutron	1.672×10^{-27}	Zero
Electron	9.109×10^{-31}	1.602×10^{-19}

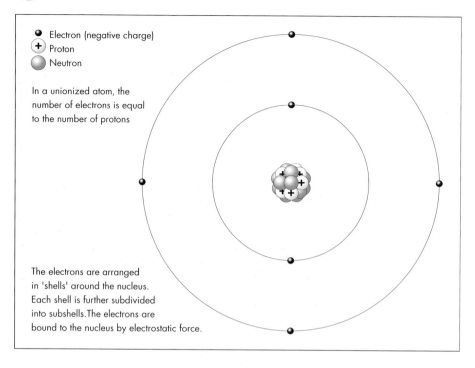

In a unionized atom, the number of electrons is equal to the number of protons

The electrons are arranged in 'shells' around the nucleus. Each shell is further subdivided into subshells. The electrons are bound to the nucleus by electrostatic force.

The number of protons in the nucleus, called the *atomic number* (or *proton number*), Z, alone determines the kind of atomic element (e.g. hydrogen: 1; carbon: 6; oxygen: 8). Also in the nucleus are neutrons, whose number usually equals or exceeds that of the protons. The combined number of protons and neutrons is called the *mass number* (or *nucleon number*), A, and is a measure of the mass of the nucleus. The nucleus of an element X can therefore be described fully in the form $^{A}X_{z}$, but the atomic number is redundant since it is specific to that chemical element, so the shorter ^{A}X is usually used.

Figure 2.1. Model of an atom.

Isotopes

It is possible for the number of neutrons to vary from atom to atom in any given element. These different forms are called *isotopes*. For example, hydrogen nuclei exist as three different isotopes: ^{1}H, ^{2}H (deuterium) and ^{3}H (tritium) (Fig. 2.2). The nuclei of different isotopes have different masses, but the same electric charge. Only very few isotopes of an atom are stable, and one of these is usually much more abundant than any of the

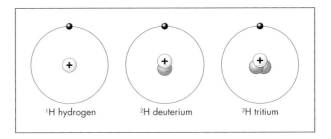

Figure 2.2. The three isotopes of hydrogen.

¹H hydrogen   ²H deuterium   ³H tritium

others; e.g. 99.99% of hydrogen is ¹H. If the number of neutrons in a nucleus varies too far from that of the most abundant isotope, the nucleus tends to become unstable and is likely to break up (fission) into smaller nuclei, a process known as *radioactive decay*. The rate of decay is measured by *half-life*, that is the time it takes for half of a sample of a radioactive isotope to decay. Some half-lives are so long that some of the element is still found naturally, such as ¹⁴C, the carbon-14 used in carbon dating, whose half-life is 5570 years.

Energy Levels

The "shells" or orbits in which the electrons normally move are at the distances from the atomic nucleus where their position is stable. The electrons are more strongly bound the nearer they are to the nucleus, so it requires energy to make them move outwards from the nucleus, and this can occur when the atoms of the gas are heated or irradiated. Thus electrons in different orbits will have different energies. Quantum theory asserts that only specific orbits and the energy levels that are associated with them are permitted and that the electron cannot exist in intervening positions or with intermediate energies. In other words, the orbits and the energy transitions are *quantized*. The important energy transitions of the electrons of a hydrogen atom are represented in Fig. 2.3. In multi-electron atoms these special representations become increasingly complex, so the transitions are usually shown diagrammatically in energy level diagrams (Fig. 2.4).

When an electron jumps from one orbital (energy level) to another, it will either emit or absorb a photon of energy. The wavelength, λ, of that photon will be determined by Planck's equation:

$$\lambda = hc/\Delta E$$

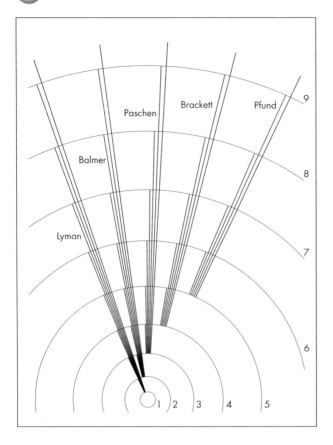

Figure 2.3. Transition series. (Developed and expanded from Koles, *Stars and their Spectra*, Fig 2.2.)

where h is Planck's constant (6.63×10^{-34} J s), c is the speed of light (2.998×10^{8} m s^{-1}) and ΔE is the change in energy level of the electron.

When one considers that the number of permitted energy levels for an electron in a hydrogen atom in interstellar space may exceed 200 (it is less in an Earth laboratory because of the relative proximity of other atoms), it becomes clear that there are a tremendously large number of possible transitions. Some manner of organization is therefore called for and the one that is used is based upon the lower energy level of the transition. An electron in its lowest possible energy state ($n = 1$) is said to be in its *ground state*. An electron anywhere else is said to be *excited*. If the electron jumps to or from its ground state ($n = 1$), the transitions form the *Lyman series* (Fig. 2.5), whose wavelengths of associated energy lie in the ultraviolet part of the spectrum. The *Balmer series* (jumps to/from $n = 2$; Fig. 2.6) lies in the visible spectrum and was therefore

Figure 2.4 (opposite). Energy level diagram for the hydrogen atom. (Developed from Zelik & Gregory, *Introductory Astronomy and Astrophysics*, Fig 8.9, and Karttunen et al. *Fundamental Astronomy*, Fig 5.2.)

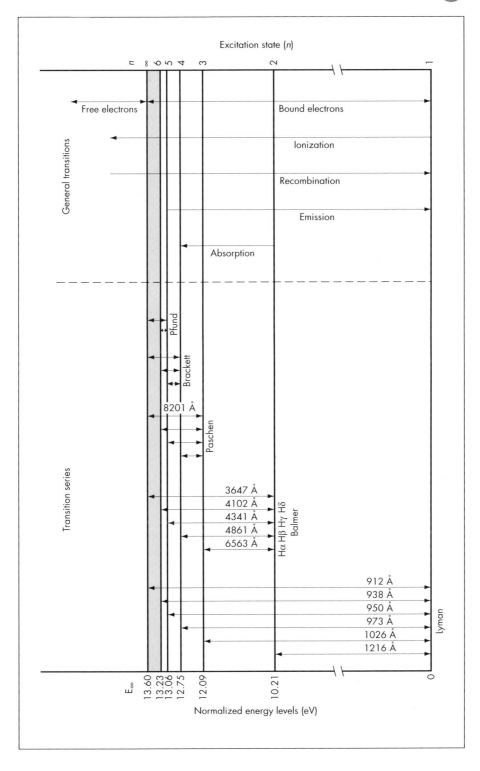

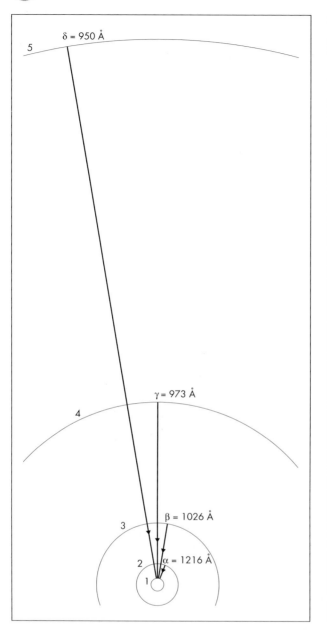

Figure 2.5. Lyman series. (Representation based on Kaler, *Stars and their Spectra*, Fig 2.2.)

the first to be studied. The *Paschen, Brackett, Pfund* and *Humphries series* (jumps to/from $n = 3$ to $n = 6$) lie in the infrared. The lowest energy jump of a series is termed α (e.g. *Lyman* α, the second β, and so on, with numbers being given from about the eighth least

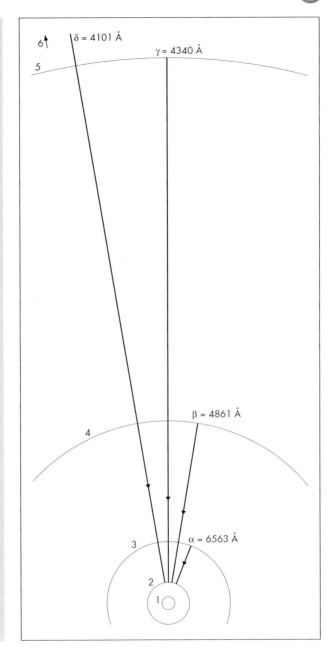

Figure 2.6. Balmer series. (Representation based on Kaler, *Stars and their Spectra*, Fig 2.2.)

energetic jump upwards (e.g. *Paschen* 12). The exception to the nomenclature is the *Balmer* series where, for historical reasons, the lowest energy jump is called hydrogen α, the next is hydrogen β, etc.

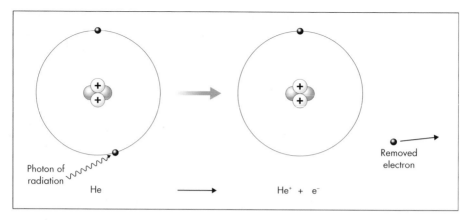

Ionization

Figure 2.7. Single ionization of a helium atom.

A normal atom has the same number of electrons as the number of protons, thus rendering the atom electrically neutral. The electrons are bound to the nucleus by their opposite electric charges. In the Bohr model of the atom, they exist in "shells" that surround the nucleus. The further the electron is from the nucleus, the less strongly it is bound to that nucleus by electrostatic attraction. All atoms or molecules in a gas are moving, their speeds increasing as the temperature of the gas increases[1] and, as a gas is heated it becomes increasingly likely that the increasingly energetic collisions between molecules will cause electrons to be stripped off the outer shells of the atoms in the molecules. The atoms will then have an imbalance of electric charge and will have a positive charge. The same removal of electrons can occur as a consequence of radiation. This process is called *ionization* (Fig. 2.7). The reverse process is called *recombination*.

The energy required to cause ionization is not quantized; neither is the energy emitted during recombination. Similarly, free electrons can change their kinetic energies (e.g. by electromagnetic interaction with each other or with other particles) and these changes need not be quantized.

The denotation of ionized atoms varies from the multiple-sign convention (He^+, Ca^{++}, Fe^{3+}, etc.) that is normally used in chemistry. The ground state of an atom is denoted by the upper case Roman numeral I, the first ionization state (loss of one electron) by II, and

---

[1]The temperature of a gas is proportional to the square of the mean speed of the molecules.

so on. Hence doubly ionized oxygen is denoted O III. The advantage of this system becomes apparent when dealing with high ionization states such as Fe XII.

The Energy Level Diagram

The energy level diagram (Fig. 2.4) can give a wealth of information about the excitation and ionization states of an atom.

Emission Spectra

When an electron jumps to a lower energy level in an atom, it emits a photon of light whose wavelength is dependent upon the energy change of the electron. Say, for example, that an electron jumps from the fourth orbit to the second orbit. If it does this directly, the photon emitted has a wavelength of 4861 Å. This is the blue-green hydrogen beta (Hβ) emission. A sufficient number of electrons making this transition in the body of a gas will result in the gas appearing to glow blue-green. If this gas was observed through a spectroscope, the observed spectrum would have only a single blue-green line. If you have not already done so, you can observe this effect by examining a distant fluorescent light source by reflection off a compact disc. The lines that you see are emission lines from the gases in the lamp, formed in a similar manner to the Hβ line described above.

It is possible, of course, that the electron might *cascade* from the fourth to the second level via the third level. In this case it would emit two photons, one in the infrared at 18751 Å, the other being the red hydrogen alpha (Hα) emission at 6563 Å (Fig. 2.8).

The energies associated with higher-level jumps are increasingly close together. Therefore their associated spectral lines will be closer together:

Hα – 6563 Å
Hβ – 4861 Å
Hγ – 4340 Å
Hδ – 4101 Å

The emission lines of hydrogen (and every other element) are never infinitely sharp, i.e. they have width and are most intense at a particular wavelength but

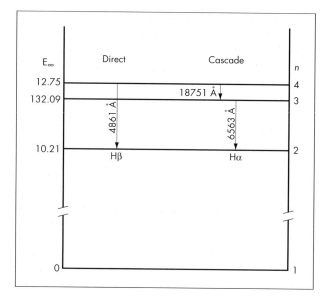

Figure 2.8. Possible transitions of an electron from level 4 to level 2.

include wavelengths within a small range either side of the primary wavelength of the line. The reasons for this spectral line broadening are:

- **Kinetic Doppler broadening.** The molecules of a gas are in constant motion, so the observed wavelengths of emission are slightly red- and blue-shifted as a consequence of any motion in the line of sight of the observer. At a temperature of 6000 K (a typical temperature of the hydrogen in the surface of a star), the 6563 Å Hα line is broadened by about 0.2 Å by the Doppler effect.

- **Quantum broadening.** The Heisenberg uncertainty principle implies that the energy of an electron in any given state is not precisely known and that the precision to which it can be known is an inverse function of the time that the electron spends in this state. Electrons typically exist for 10^{-8} s in an excited state and the effect of this is a broadening of the order of 10^{-5} Å for visible frequencies.

- **Collision broadening.** The energy levels, and therefore the spectral lines, are shifted as a consequence of perturbation by other particles. This is particularly pronounced when those particles are charged (the Stark effect[2]), so the broadening due to this will be

---

[2]The splitting of energy levels by a strong electric field. It is the electrical equivalent of the Zeeman effect.

greater when the emitting gas is ionized and, because the perturbation will increase as the proximity of particles increases, will be greater as the density (and therefore pressure) of the gas increases.

- **Zeeman effect broadening.** Magnetic fields interact with the magnetic dipole of electrons and cause the energy level to split into multiple components. If the magnetic field is sufficiently strong and uniform (e.g. in sunspots), the splits in the resulting spectral lines may be resolvable (the Paschen–Back effect). This permits deductions about the nature of the magnetic field.

- **Other Doppler broadening.** Unless a star's spin axis is along the observer's line of site, one limb will be approaching us, the other receding. The effect of this is a red and blue shifting of the spectral lines, which broadens them. Similarly, any turbulence or pulsation in the surface of a star will shift the lines and cause them to broaden.

Continuous Spectra

The broadening of spectral lines can cause them to overlap, resulting in a continuous spectrum. Thus hot dense gases, where collision broadening is significant, may produce continuous spectra, as will hot liquids and solids. The other important mechanisms for the production of continuous spectra involve free electrons. The energy change during recombination is not quantized, so neither will the wavelengths of the emitted photons of *free-bound* interactions. The continuous spectra that result will show greatest intensities at the wavelength limits of the transition series (e.g. 911.8 Å for the hydrogen Lyman series). The energy changes during *free-free* interactions are also not quantized, so continuous spectra may result from these interactions.

Absorption Spectra

If continuous radiation passes through a cool transparent gas (Fig. 2.9), those photons whose energies exactly match the energy differences between any two

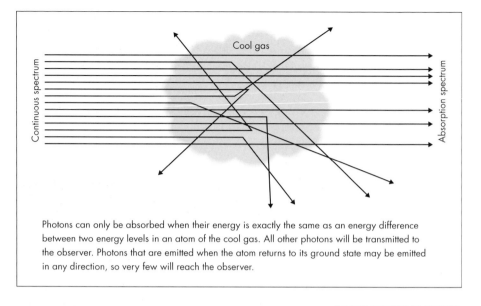

Photons can only be absorbed when their energy is exactly the same as an energy difference between two energy levels in an atom of the cool gas. All other photons will be transmitted to the observer. Photons that are emitted when the atom returns to its ground state may be emitted in any direction, so very few will reach the observer.

Figure 2.9.
Formation of an absorption spectrum.

excitation levels in an atom of the gas may be absorbed. The resulting spectrum will then have dark lines at the wavelengths corresponding to those energies.[3] Given that the typical period for atoms to exist in their excited states is of the order of 10^{-8} s, it is reasonable to wonder how absorption spectra can occur, given that the atom will emit a photon immediately after it has absorbed one. The emitted photons may travel in any direction, so very few will travel in the line of sight towards an observer. It is also possible that the electron will not return directly to its ground state but may cascade through other levels (see above). In this case the photons emitted will be of lower energies (i.e. longer wavelengths) than the absorbed photon and so cannot illuminate the dark absorption line no matter which direction they travel.

Kirchhoff's Laws

In the 19th century Gustav Kirchhoff encapsulated the basic principles relating the appearance of spectra to their causes into three empirical laws:

1. An incandescent solid or liquid or a gas under high pressure will produce a continuous spectrum.

---

[3]These lines are, of course, in exactly the same place as the emission line corresponding to the transition of the electron.

2. An incandescent gas under low pressure will produce an emission-line spectrum. The position of these lines will depend upon the elements present in the gas.

3. A continuous spectrum viewed through a transparent low-density gas at low temperature will produce an absorption-line spectrum. The position of these lines will depend upon the elements present in the cool gas.

Other Atoms, Ions and Molecules

The picture of emission spectra given above for the hydrogen atom can become considerably more complex in the atoms of other elements, where the electrons in the atoms interact with each other. A consequence of this is that no two elements produce the same spectral signature, potentially allowing the chemical analysis of a gas by examination of the spectral lines that are present.

The Pauli exclusion principle implies that only two electrons may occupy the innermost shell, eight the second shell, and so on. "Surplus" electrons occupy incomplete outer shells. Filled shells are tightly bound to the nucleus and have a shielding effect between the nucleus and the outer electrons, so these are less tightly bound. These are the electrons, known as *valence electrons*, that are largely responsible for the chemical nature of an element; because of this looser binding, they are easily excited and the atom is easily ionized. The spectrum of lithium, which has one valence electron outside one filled shell, is similar to that of hydrogen but has much lower ionization and excitation potentials and thus the Lyman series limit is at a longer wavelength (2250 Å as compared to 912 Å).

Where there is more than one valence electron, the situation is significantly more complex, as are the resulting spectra.

Ionization will also modify a spectrum. Taking the simple case of helium, if it becomes singly ionized to He^+ (He II), it has one remaining electron, but because it is bound to the nucleus by the electrostatic attraction of two protons as compared to the one proton in a hydrogen atom, the energy associated with any equivalent transition will be greater[4] and the equivalent spectral lines will be shifted to the blue/violet end of the spectrum. In the case of the He II Balmer series, they will move from the visible part of the spectrum into that occupied by the hydrogen Lyman series. Similarly, the He II Paschen and Bracket series will shift into the visible spectrum.

Forbidden Lines

In addition to telling us what energy transitions are possible in an atom, the quantum conditions also tell us the probability of occurrence of any given transition. The more probable transitions will result in stronger lines. There are some states of an electron in an atom, the *metastable states*, from which transitions are extremely unlikely. *Forbidden lines* are the spectral lines that would result from a transition from a metastable state. "Forbidden" is a misnomer as the transitions are not forbidden but merely very improbable.

Figure 2.10 shows the five lowest energy states for O III. The levels $n = 3$ and $n = 4$ represent metastable states and transitions from these states are not observed in a laboratory; they do occur, but in insufficient quantities to produce an observable line. For a large gaseous nebula (e.g. the Great Orion Nebula), any spectral analysis will involve a tremendous volume of gas, and hence of O III atoms, and there are sufficient transitions to render the corresponding spectral lines visible.

In the rarefied environment of a gaseous nebulae, there is often inadequate excitation to energize an electron to one of the higher energy states that would give rise to the permitted lines. The permitted lines are therefore relatively weak by comparison with the forbidden lines.

By convention, the forbidden lines in a spectrum are denoted by square brackets, e.g. [O III].

Doppler Effect; Red- and Blue-shift

The Doppler effect is the apparent change of wavelength (or frequency) of wave energy resulting from the

---

[4]In this instance of two protons, it will increase by a factor of 4 (2^2).

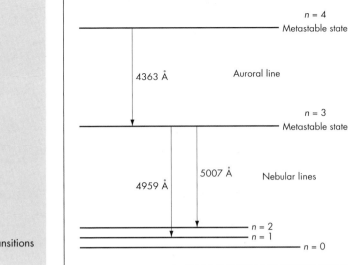

Figure 2.10.
"Forbidden" transitions
of O III.

relative motion of the source and the observer. When the source is moving towards the observer, the waves become "squashed" and the wavelength decreases. When the source is receding from the observer, the waves become "stretched" and the wavelength increases (Fig. 2.11).

If the wavelength of visible light decreases, the light will appear bluer than it would if the source had had no motion in the observer's line of sight. This is *blue shift*. If the wavelength of visible light increases, the light will appear redder than it would if the source had had no motion in the observer's line of sight. This is *red shift*.

For electromagnetic waves, the observed wavelength, λ_0, is given by:

$$\lambda_0 = \frac{1 + v/c}{\sqrt{1 - v/c}} \lambda_s \qquad (2.1)$$

where λ_s is the wavelength of light emitted by the source, v is the relative line of sight velocity of the source with respect to the observer, and c is the speed of light. If the spectrum of a star is compared with the spectrum of a stationary source, the change in position of spectral lines (see Fig. 2.12) then indicates the relative line of sight velocity of the moving source in accordance with Equation (2.1) above.

It is important to note that, in addition to Doppler red shift, red shift will result from strong gravitational fields (gravitational red shift) and the expansion of space (cosmological red shift).

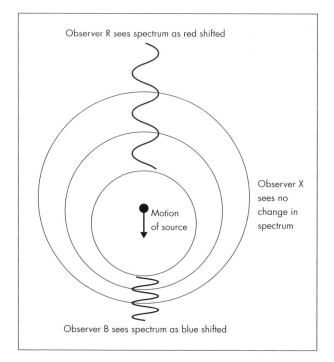

Observer R sees spectrum as red shifted

Motion
of source

Observer X
sees no
change in
spectrum

Observer B sees spectrum as blue shifted

Figure 2.11. Doppler effect.

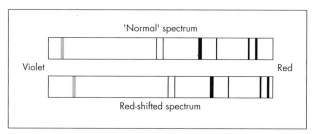

'Normal' spectrum

Violet

Red

Red-shifted spectrum

Figure 2.12. Red shift of spectral lines.

Bibliography and References

Duncan, T (1982) *Physics*. John Murray, London, ISBN 0719538890.

Karttunen, H et al. (1996) *Fundamental Astronomy*, 3rd edn. Springer-Verlag, Heidelberg, ISBN 3540609369.

Payne-Gaposchkin, C (1956) *Introduction to Astronomy*. University paperbacks, London.

Ridpath, I (1997) *A Dictionary of Astronomy*. Oxford University Press, ISBN 0192115960.

Zelik, M and Gregory, S (1998) *Introductory Astronomy and Astrophysics*, 4th edn. Saunders College Publishing, Fort Worth, ISBN 0030062284.

Chapter 3

Spectroscopes

Stephen Tonkin

Dispersion of Light

Spectroscopes disperse light using either prisms or diffraction gratings. Early spectrometers used prisms. When light passes across a boundary between two materials of different refractive indices, n_1 and n_2, it is refracted in accordance with *Snell's law*,

$$n_1 \sin i_1 = n_2 \sin i_2,$$

where i_1 and i_2 are the angles of the incident and refracted rays respectively. The dispersion in a prism results from the different relative refractive indices, $n = (n_2/n_1)$, of the prism material for different wavelengths of light (Fig. 3.1). The longer wavelengths (red end of the spectrum) are refracted less than the shorter ones (blue end). A clean spectrum (i.e. one in which the colors do not overlap) is obtained by passing the light through a slit, then collimating it with a lens. A second lens focuses the dispersed light and the resulting spectrum is essentially multiple images of the slit.

Prisms have the disadvantage that the dispersion is nonlinear. The dispersion is less at the red end of the spectrum, so prisms are especially unsuitable for studying spectral characteristics at long wavelengths.

Objective Prisms

One of the simplest way of obtaining stellar spectra is to place a prism in front of the aperture of a telescope

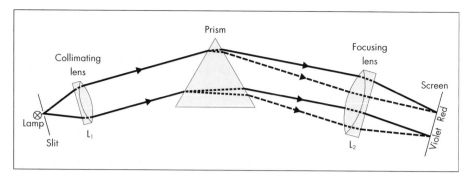

or camera (Fig. 3.2). Large glass prisms are both expensive and difficult to obtain, but liquid-filled prisms are relatively simple to construct. The light from a point source at a distance can be considered to be equivalent to collimated light from a slit, and a clean spectrum can thus be obtained, with the objective lens of the telescope being equivalent to the lens L_2 in Fig. 3.1.

Figure 3.1. The dispersion of white light by a prism.

Direct Vision Prism Spectroscopes

An arrangement of an odd number of prisms of materials with two different refractive indices is called an Amici prism.[1] It can be designed to disperse light without either deviating or displacing it (Fig. 3.3) and can thus be used as a direct vision spectroscope.

Figure 3.2. Objective prism.

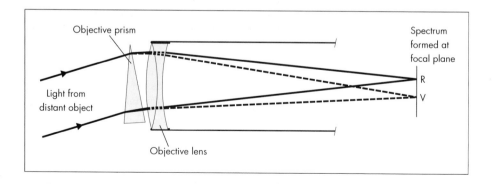

[1]This should not be confused with the roof prism of the same name that is used to give erect images in astronomical telescopes.

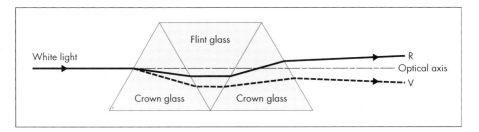

Figure 3.3. Amici prism.

Diffraction Gratings

Most modern spectroscopes use diffraction gratings. There are two basic forms: transmission gratings and reflection gratings. A modern transmission grating consists of closely spaced grooves in a substrate, usually glass. Either the grooves act as centers of scattering of the incident light (amplitude grating) or the varying optical thickness of the substrate causes phase changes (phase grating). A reflection grating has grooves ruled in a reflective substrate. The mathematical treatment of transmission and reflection gratings is similar. The spacing of slits or grooves is given in lines/mm.

Diffraction is a phenomenon based on the wave nature of light. If we consider diffraction of light by two slits (Fig. 3.4), it is apparent that there are regions where there is constructive interference and regions where the interference is destructive. The constructive interference occurs where the path difference between the light from the two slits is an integer number of wavelengths of light. Hence constructive interference will be at a different angle for different colors (wavelengths). Adding further slits increases the interference effect and thus the color separation.

An algebraic treatment for a normal incident ray (Fig. 3.5) yields the *diffraction grating formula* for constructive interference:

$$n\lambda = a \sin\theta_n$$

where n is the order of diffraction, λ is the wavelength of the light, a is the slit separation and θ_n is the angle between the direction of the nth-order and the normal to the grating.

Generalizing for a nonnormal incident ray at an angle of θ_i to the normal (Fig. 3.6) yields the formula:

$$n\lambda = a \sin(\theta_i = \theta_n).$$

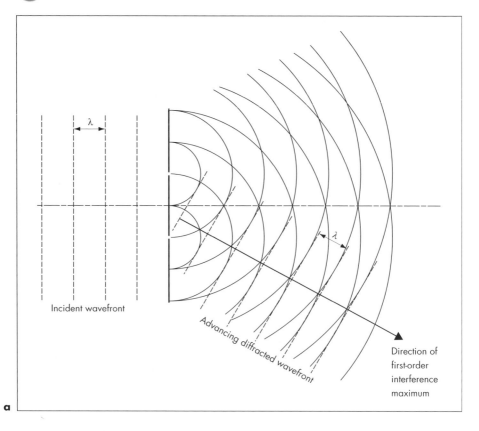

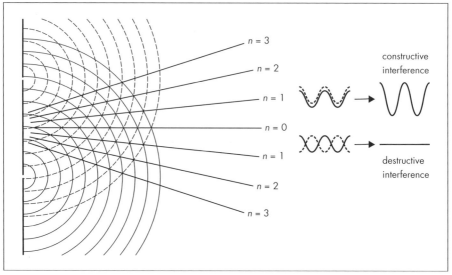

Figure 3.4. Diffraction from two slits.

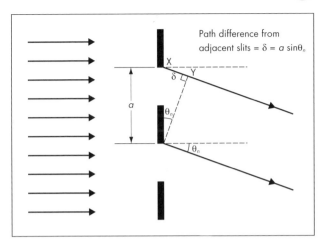

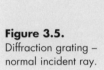

Figure 3.5.
Diffraction grating –
normal incident ray.

A sign convention is used such that when the diffracted ray is on the opposite side of the normal to the incident ray, θ_n is signed negative, as is the order of the diffraction, n.

Application of the grating formula to the wavelength range of the light being diffracted will show the spread of each order of the spectrum and any overlap that exists. Figure 3.7 shows this graphically for a source with a spectral range of 4000–7000 Å and a grating of 600 lines/mm. It shows how higher orders of spectra are more greatly dispersed but that they are also more subject to overlap. This is because $n\lambda$ can have the same value for different wavelengths in different orders of spectra. The nonoverlapping

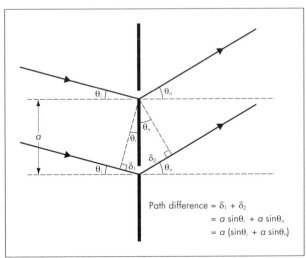

Figure 3.6.
Diffraction grating –
non-normal incident ray.

portion of any order is known as the *free spectral range* of that order. If λ_1 is the shortest wavelength in a spectrum, the free spectral range, F, of any order of spectrum is given by:

$$F = \lambda_1/n.$$

Dispersion and Resolution

Figure 3.7 shows how higher orders of spectra are more dispersed. They are also less intense. In practice, a lens or mirror will focus the spectrum produced by the grating on to the detector (screen, photographic film, CCD). The *linear dispersion, D,* of such a system is given by:

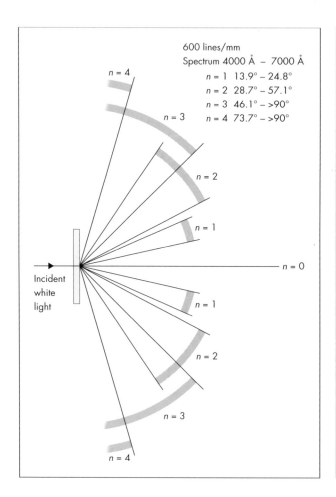

600 lines/mm
Spectrum 4000 Å – 7000 Å
$n = 1$ 13.9° – 24.8°
$n = 2$ 28.7° – 57.1°
$n = 3$ 46.1° – >90°
$n = 4$ 73.7° – >90°

Figure 3.7. Spread of spectra from a diffraction grating – normal incidence.

$$D = \frac{fn}{a cos\theta_n}$$

where f is the focal length of the lens or mirror.

As indicated above, increasing the number of grooves in a grating increases the sharpness and intensity of the spectral lines. The ability of a grating to enable the spectral lines due to two close wavelengths to be distinguished is called its *resolution*, R, and is a function of both the width (W) of the grating and its line density.[2] It is given by:

$$R = \frac{W sin\theta_n}{\lambda}$$

Blazing

Figure 3.8.
Diffraction at a single slit.

A single slit will also give rise to diffraction (Fig. 3.8) and the greatest light intensity (Fig. 3.9a) exists when the path difference from different parts of the slit is zero ($\beta = 0$). This is the direction in which the ray would be

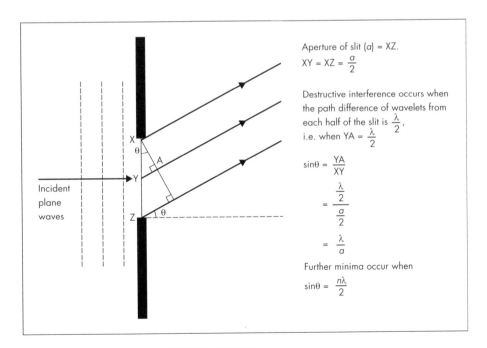

Aperture of slit (a) = XZ.

$XY = XZ = \frac{a}{2}$

Destructive interference occurs when the path difference of wavelets from each half of the slit is $\frac{\lambda}{2}$, i.e. when $YA = \frac{\lambda}{2}$

$sin\theta = \frac{YA}{XY}$

$= \frac{\frac{\lambda}{2}}{\frac{a}{2}}$

$= \frac{\lambda}{a}$

Further minima occur when

$sin\theta = \frac{n\lambda}{2}$

Incident plane waves

---

[2]The effect of line density is not immediately apparent from this equation, which is derived from R = nN, where N is the number of lines. $N = W/a$; $n = (sin\theta_n)/\lambda$.

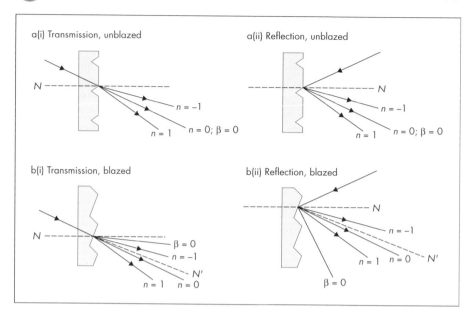

expected to pass as a consequence of geometrical optics. In normal gratings this is coincident with the zeroth-order spectrum, in which there is no dispersion.

Blazing is a technique of angling the surface of the grooves in a grating so that it shifts the position of maximum intensity, $\beta = 0$, from the zeroth-order spectrum to a higher-order spectrum, thus rendering the higher order brighter. This is shown in Fig. 3.9b. N' is the normal to the surface of the groove. In the transmission grating, $\beta = 0$ is shifted to its direction to which the incident ray would be refracted in accordance with Snell's law. In the reflection grating, $\beta = 0$ is shifted to its direction to which the incident ray would be refracted in accordance with the laws of reflection in geometrical optics.

Figure 3.9. The effect of blazing.

Objective Gratings

Among the simplest ways of obtaining spectra is that of placing a transmission grating over the aperture of a telescope or camera. This grating can be as simple as a net curtain. Alternatively, Bev Ewen-Smith of COAA has produced some free software (Windows 3.x and 9x) to enable an objective grating to be printed onto a transparency (http://www.ip.pt/coaa/software.htm). The reflection grating equivalent of these methods uses CDs or DVDs and is described in Part 2 by David Randell.

The spectra resulting from an objective grating (or prism) are focused at the focal point of the camera or telescope. If they are recorded on photographic film, the spectrum of every star in the field will be recorded, provided they are sufficiently bright. The spectral lines will be more easily visible if the stars are permitted to drift slightly across the field of view in a direction perpendicular to the direction of dispersion of the grating. This is most easily achieved for bright stars by adjusting the grating so that the dispersion is in the plane of the spectroscopist's meridian, and allowing Earth's rotation to provide the drift.

Star Spectroscopes

The "star spectroscope" (a.k.a. "eyepiece" spectroscope) is a small direct vision transmission grating that is placed in the barrel of an eyepiece. The most readily available example is the *Rainbow Optics Star Spectroscope* whose use is described by Jack Martin in Part 2. It consists of a blazed grating[3] of about 600 lines/mm that screws into the thread of a standard 1.25 inch (31.7 mm) eyepiece barrel and a cylindrical lens that fits over the ocular end of the eyepiece. This latter lens widens the spectrum, which would otherwise be a thin line, enabling spectral lines to be seen. There is no need for a slit as the intended spectral sources, stars, are point objects. For extended objects, such as the Sun or Moon, a slit needs to be introduced. One "Heath-Robinson" way around this for solar spectra is to use a polished cylinder (such as part of a telescopic antenna for a radio) at a distance and align the telescope to that; the reflection from the cylinder is a line and is optically equivalent to a slit.

Slit Spectrometers[4]

Most amateur (and professional!) spectroscopists use slit spectrometers, a number of which are described in Section 2.

---

[3] The *Rainbow Optics* grating is blazed so that 75% of the light energy goes into one of the first-order spectra.

[4] The words spectroscope, spectrograph and spectrometer are used interchangeably throughout this book. Strictly, a spectro*scope* uses the eye as the detector, a spectro*graph* uses photographic film, and a spectro*meter* uses an electronic detector (usually a CCD in amateur instruments).

The essential components of a slit spectrometer are:

- slit
- collimating lens or mirror
- grating (usually reflection)
- focusing mirror or lens
- detector.

The collimating and focusing elements may be combined, as in the *Littrow Spectrograph* in Fig. 3.10.

The Czerny–Turner Spectrometer whose optical schematic is shown in Fig. 3.11 typifies a more complex arrangement. Light from the entrance slit is collimated by the collimating mirror, which directs it on to the grating. The spectrum from the grating is focused by the second mirror through the exit slit. Only a small part of the spectrum, dependent upon the width of the slit, is visible to the detector. The grating can be rotated to permit other regions of the spectrum to be examined.

Modern amateur slit spectrometers are frequently fiber-fed. The telescope focuses the star on to a cylindrical bundle of optical fibers. The other end of the fiber bundle has the fibers arranged in a line (Fig. 3.12). This has the

Figure 3.10. Littrow spectrograph.

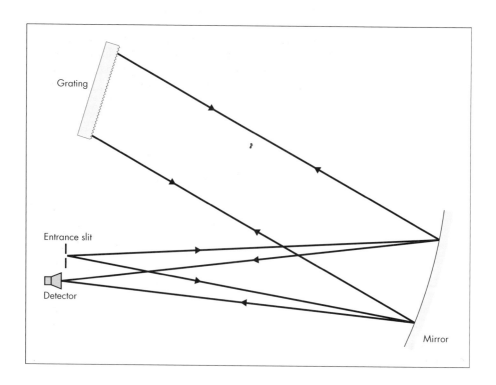

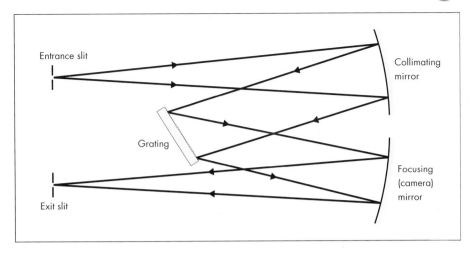

Figure 3.11.
Czerny–Turner
spectrograph.

advantages of creating a slit (the line of fibers), permitting the light from other fibers to be used for autotracking the star, allowing fibers from reference spectral sources to be fed to either end of the fiber "slit", and permitting the environment of the spectrometer to be more easily maintained constant. This last is particularly important, as spectrometers are very sensitive to changes in temperature, which can throw them out of calibration. Amongst the disadvantages is the loss of light in the fibers.

The recorded spectrum is multiple images of the spectrometer's slit, each at a different wavelength. The slit also has the effect of greatly attenuating optical "noise" such as skyglow, thus permitting amateur spectrometry to be undertaken from urban environments.

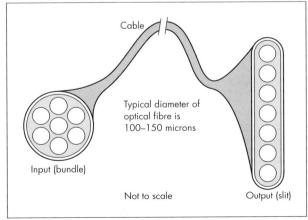

Figure 3.12. Optical fiber feed.

Calibration

A spectrometer must be calibrated both for wavelength and to account for the different sensitivity of the detector at different wavelengths (*flux* calibration). The usual reference sources for wavelength calibration are mercury and neon lamps. As well as being commonly available, together they offer a wide range of spectral lines.[5] An alternative to lamps is to calibrate against the spectra of stars whose spectral characteristics are known. Flux calibration is normally done with reference to standard spectrometric stars.

Replica Gratings

Many of the gratings available to amateurs are termed *replica gratings*. Master gratings are very expensive to produce, so replicas are produced from the master[6] by a process similar to molding. The replica is mounted on to a blank of glass or silica and, if it is to be a reflection grating, is aluminized in a vacuum chamber. Good modern replica gratings can be at least as good as the original masters; some are better.

Bibliography and References

Duncan, T (1982) *Physics.* John Murray, London, ISBN 0719538890.
Pedrotti, F and Pedrotti, T (1993) *Introduction to Optics*, 2nd edn. Prentice-Hall, Englewood Cliffs, ISBN 0130169730.
Sidgwick, J B (1971) *Amateur Astronomer's Handbook*, 3rd edn. Dover, New York, ISBN 0486240347.

---

[5]Many portable fluorescent lamps are mercury vapor lamps; neon lamps are available as mains electricity indicator lamps.
[6]In practice, a good replica is usually used as a "submaster" and further replicas are made from that.

Chapter 4
The Spectral Classification of Stars

Stephen Tonkin

Chapter 1 introduced the beginnings of the spectral classification that is used today. Its main characteristics are:

- Twelve spectral types, denoted by upper-case Roman letters:

$$
\begin{array}{c}
\text{R}-\text{N} \\
\diagup \\
\text{W}-\text{O}-\text{B}-\text{A}-\text{F}-\text{G}-\text{K}-\text{M} \\
\diagdown \quad \diagdown \\
\text{C} \quad \text{S}
\end{array}
$$

The OBAFGKM sequence is based on the Harvard College Observatory system due to Cannon.[1] W was added to account for Wolf–Rayet stars (further subdivided into WC and WN, with strong carbon and nitrogen lines respectively.) The R–N side-branch from type G has orange and red stars (recall that G stars are yellow) and the C side-branch stars contain a great deal of carbon. The S side-branch of K are giant long-period variables with bands of TiO and ZrO.

The spectrum of a star is related to its surface temperature, with hotter stars having spectra more to the left of the spectral classification above. Owing to an early misconception about stellar evolution, stars with spectra at the OBA end are sometimes

---

[1]See p.6.

referred to as "early", and those at the GKM end as "late".

- Decimal subdivisions into classes, denoted by Arabic numerals (0, 1,. .., 9), also based on Cannon's system. The exceptions are types W and O; classes 0–4 of type O are now classes W5–W9.

- Suffixes, denoted by lower-case Roman letters, describing the spectral lines:

  e   emission lines
  em   metal emission lines
  k   absorption lines of interstellar calcium
  m   strong metallic absorption lines
  n   nebulous lines (due to spinning star)
  nn   very nebulous lines
  p   peculiar spectrum
  s   sharp lines
  ss   very sharp lines
  v   variation in the spectrum
  w   wide lines
  wk   weak lines

- Suffixes or prefixes describing other aspects of the source:

  MSx multiple star system ("x" denotes the number of components)
  SB   spectroscopic binary
  var variable star
  Chemical symbol, e.g. Ca prominent lines due to that element, e.g. calcium.

- Six luminosity classes, denoted by upper-case Roman numerals and lower-case Roman letters. This classification is called the *MKK* (or *Yerkes*) classification and is due to William Morgan, Philip Keenan and Edith Kellman of Yerkes Observatory.

  Ia   most luminous supergiants
  Ib   less luminous supergiants
  II   luminous giants
  III ordinary giants
  IV   subgiants
  V   main sequence stars (a.k.a. dwarfs)

  A seventh class, VI (or sd), is sometimes used to describe small (red) dwarfs and an eighth, VII (or wd or D) to describe white dwarfs. A "?" indicates that the luminosity class is uncertain.

Using this classification, the variable red giant star Mira (o Cet), for example, is described as M7e III.

Magnitudes and Color Index

The *apparent magnitude*, *m*, of a star is a measure of its perceived brightness and is based on a scale, introduced by Hipparchus in the 2nd century BCE, in which the first stars to be seen at twilight (first magnitude stars) are about 100 times brighter than the faintest stars visible to the naked eye (sixth magnitude). A lesser magnitude number signifies a brighter star. This scale was formalized mathematically by N.R. Pogson some 2000 years after Hipparchus, such that a difference of 5 magnitudes is exactly a hundredfold difference in brightness, so that a difference of one magnitude is a difference of $\sqrt[5]{100} = 2.512$. Apparent magnitude is a function of the luminosity and distance of the star and the transparency of the medium through which it is viewed.

The measure of the luminosity of a star is its *absolute magnitude*, *M*. It is the apparent magnitude of the star as it would appear at a distance of 10 parsecs (32.6 light years) if viewed through a perfectly transparent medium.

The distance modulus is the difference between the apparent and absolute magnitudes of the star, $m - M$. It is related to the distance, *D*, by the equation:

$$m - M = 5 \log D - 5.$$

Because eyes and cameras (film and electronic) are differently sensitive to light, photographic and visual magnitudes of the same object usually differ. Photographic emulsion is more sensitive in the blue; the eye is more sensitive in the yellow. The color index (CI) of a star is the difference between its photographic magnitude, B, measured at 4259 Å (blue) and its photovisual magnitude, V, measured at 5280 Å (yellow). Hence CI = B − V. Hotter stars radiate more strongly at the blue region of the spectrum,[2] thus hotter stars have a negative CI and cooler ones have a positive CI.[3]

There are several variations on this theme, such as the UBV system (ultraviolet at 3600 Å, blue at 4200 Å, V at 5400 Å, or the UVBRI system that incorporates

---

[2] Wien's displacement law. It applies to perfect blackbodies, but is a reasonable approximation for most celestial objects.
[3] Recall that brighter stars have a lower magnitude!

magnitudes in the red and infrared. The zero point on the CI scale is where $U - B = B - V = 0$ and is where the surface temperature of the star is 10 000 K, i.e. stars of spectral type A0 (e.g. α Lyr).

The Hertzsprung–Russell Diagram

The Hertzsprung–Russell diagram (Fig. 4.1) is a two-dimensional plot of the properties of stars in which spectral type (equivalent to surface temperature) is plotted against absolute visual magnitude (equivalent to luminance). The majority of stars fall on a sigmoid curve that is called the *main sequence* (because that is where most of the stars lie). Stars of the same spectral type can have different absolute magnitudes. This is because stars are not all of the same size and larger stars have larger surfaces from which they radiate and are thus brighter. There are several bands of stars on the HR diagram that are not on the main sequence. These are the dwarfs (below the main sequence) and the giants and supergiants (above the main sequence).

A similar diagram is produced if color index is plotted against luminosity.

The characteristics of each spectral class are given in the following tables.

Bibliography and References

Kaler, J (1997) *Stars and Their Spectra*. Cambridge University Press, ISBN 0521585708.

Karttunen, H et al. (1996) *Fundamental Astronomy*, 3rd edn. Springer-Verlag, Heidelberg, ISBN 3540609369.

Malpas, B. *Catalogue of Bright Star Spectra*. http://users.erols.com/njastro/faas/pages/starcat.htm

Payne-Gaposchkin, C (1956) *Introduction to Astronomy*. University paperbacks, London.

Van Zyl, J (1996) *Unveiling the Universe*. Springer-Verlag, London, ISBN 3540760237.

Zelik, M and Gregory, S (1998) *Introductory Astronomy and Astrophysics*, 4th edn. Saunders College Publishing, Fort Worth, ISBN 0030062284.

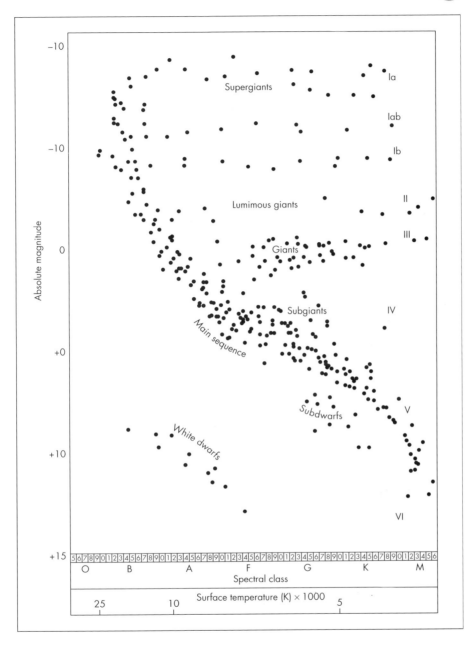

Figure 4.1. The Hertzsprung–Russell diagram.

Class W

Temperature	20 000 K to 80 000 K
Prominent spectral lines	N III 4100 Å (WN stars) He II 4340 Å He I 4471 Å N III 4340 Å (WN stars) C III, C IV 4650 (WC stars) He II 4686 Å He II 4861 Å
Examples	γ Vel (WC8), EZ CMa (WN5), HD 16523 (WC6)
Comments	Wolf-Rayet stars These are the stars at the hearts of planetary nebulae WC: carbon emission bands WN: nitrogen emission bands Classified as type O (and sometimes B) in older catalogs

Class O

Temperature 25 000 K to 40 000 K

Prominent spectral lines
Si V 4089 Å
N III 4097 Å
Hγ 4340 Å
He I 4471 Å
He II 4542 Å
C III 4650 Å

Examples σ Ori (O5V), θ¹ Ori (O6pe), ξ Per (O7Iab), γ² Vel (O7.5 I), δ Ori (O9II), ζ Ori (O9 I bSB), ζ Pup (O5Iaf).

Spectrum of
δ Ori O9
30 000 K

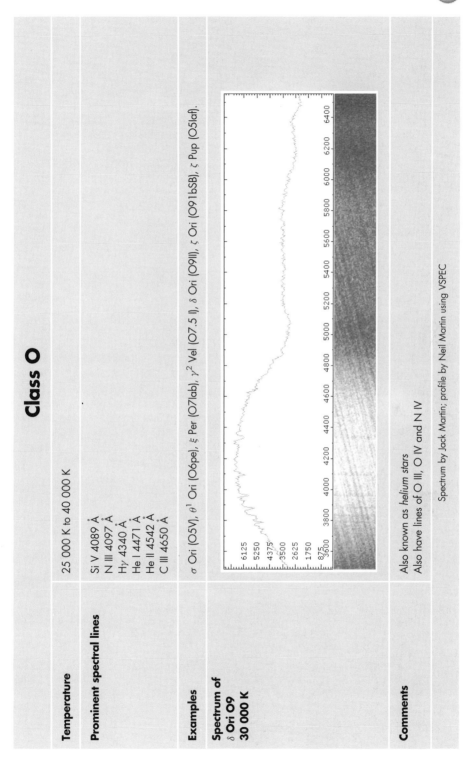

Spectrum by Jack Martin; profile by Neil Martin using VSPEC

Comments
Also known as *helium* stars
Also have lines of O III, O IV and N IV

Class B

Temperature 11 000 K to 25 000 K

Prominent spectral lines

Si IV 4089 Å
Hδ 4102 Å
Si II 4128 Å
He I 4471 Å
C III 4540 Å
He II 4541 Å

Examples γ Cas (B0IVe), ε Ori (B0Ia), α Vir (B1V), γ Ori (B2III), η UMa (B3V SB), β Ori (B8Ia).

Spectrum of
γ Ori
B2
20 000 K

Comments

Great variety of spectra requires subdivision of decimal classes
Sometimes have hydrogen emission lines (Be stars)
No ionized helium
O II and N II lines become fainter from B0 to B9
All the naked-eye Pleiades are class B
Sanduleak-69° 202 (which became SN1987a) was a B3Iab star

Spectrum by Jack Martin; profile by Neil Martin using VSPEC

Class A

Temperature	7500 K to 11 000 K
Prominent spectral lines	Hε 3970 Å
	Hδ 4102 Å
	Fe I 4299 Å
	Ti II 4303 Å
	Hγ 4340 Å
	Fe I 4383 Å
	Hβ 4861 Å
	Mg II 5173 Å
	Hα 6563 Å

Examples

α CMa(A0 Vm), α Lyr (A0 Vvar), α Cyg (A2 Ia), α Gem (A2 Vm), α PsA (A3 V), α Aql (A7 IV-V)

Spectrum of
α Gem
A1
9500 K

Spectrum by Jack Martin; profile by Neil Martin using VSPEC

Comments

These hydrogen lines are the *Balmer series*
No helium lines
H absorption lines strongest in A1 and A2
The Ca II lines 3934 Å (H line) and 3968 Å (K line) become apparent

Class F

Temperature

6000 K to 7500 K

Prominent spectral lines

Ca II 3934Å (H line)
Ca II 3968Å (K line)
Hε 3970 Å
Hδ 4102 Å

Examples

α Car (F0Ia), α CMi (F5IV), δ CMi (F8Ia)

Spectrum of
β Cas
F2
7100 K

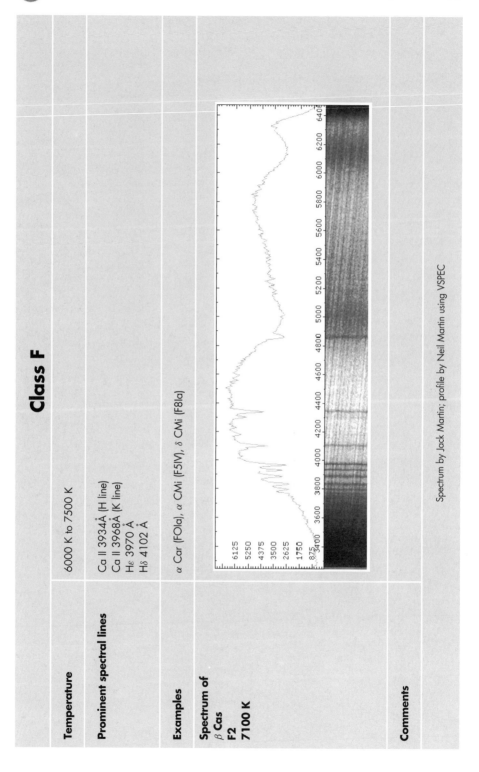

Comments

Class G

Temperature

5000 K to 6000 K

Prominent spectral lines

Examples

Sun (G2V), α Aur (G8III)

Spectrum of
α Aur
G8
5300 K

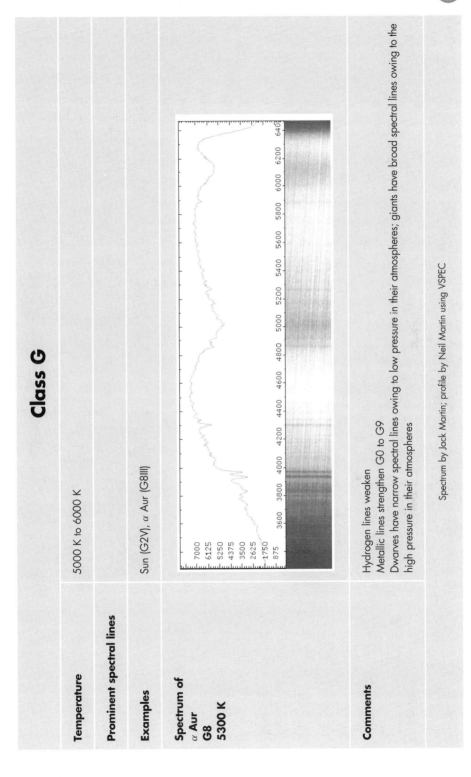

Comments

Hydrogen lines weaken

Metallic lines strengthen G0 to G9

Dwarfs have narrow spectral lines owing to low pressure in their atmospheres; giants have broad spectral lines owing to the high pressure in their atmospheres

Spectrum by Jack Martin; profile by Neil Martin using VSPEC

Class K

Temperature
Prominent spectral lines

3500 K to 5000 K
Ca II 3934 Å (H line)
Ca II 3968 Å (K line)
Fe III 4172 Å
Fe I 4325 Å
Hγ 4340 Å
Fe I 4383 Å

Examples

ε Eri (K2V), α Boo (K2III), α Tau (K5III)

Spectrum of
α Boo
K2
4830 K

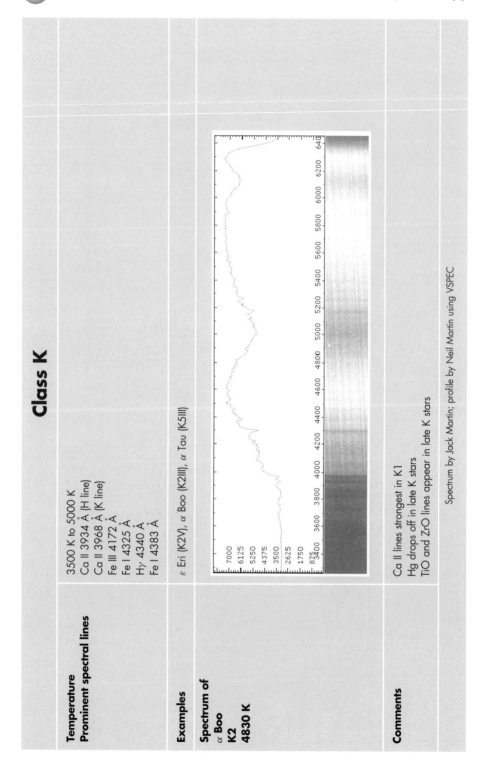

Spectrum by Jack Martin; profile by Neil Martin using VSPEC

Comments

Ca II lines strongest in K1
Hg drops off in late K stars
TiO and ZrO lines appear in late K stars

Class M

Temperature

3000 K to 3500 K

Prominent spectral lines

Fluted molecular bands of TiO

Examples

α Sco (M1 Ib), α Ori (M2 Iab), β Peg (M2 II var), γ Cru (M3 III)

**Spectrum of
β Peg
M2
3400 K**

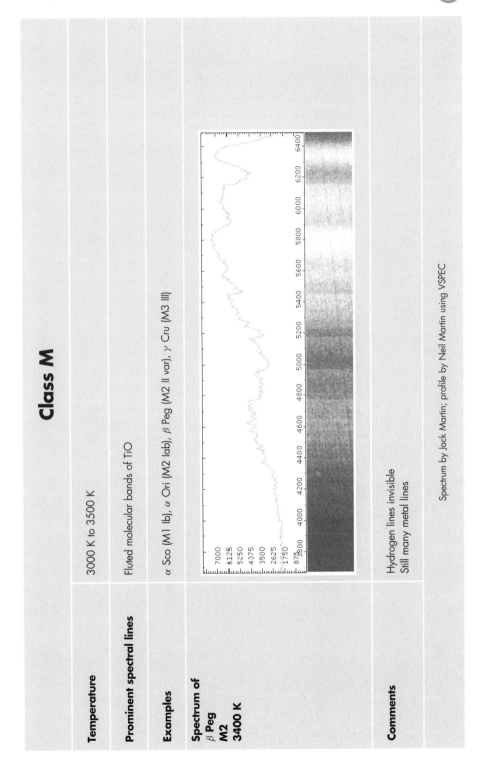

Comments

Hydrogen lines invisible
Still many metal lines

Spectrum by Jack Martin; profile by Neil Martin using VSPEC

Class R

Temperature	2000 K to 3000 K
Prominent spectral lines	Fluted carbon bands
Examples	UV Cam (R8v), NQ Gem (R9)
Comments	Bright in violet and blue Classification is difficult for these stars, hence many R-type stars are also classified as C

Class N

Temperature	2000 K to 3000 K
Prominent spectral lines	Fluted carbon bands
Examples	ST Cam (N5), W Ori (N5)
Comments	Irregular variables Very red Classification is difficult for these stars, hence many N-type stars are also classified as C

Class C

Temperature	2000 K to 3000 K
Prominent spectral lines	Very strong carbon
Examples	R CrB (C0; also classified as F8pe), W CMa (C5II), V Ari (C5II), SU And (C5 II), W Cas (C0ev), R CMi (C0ev)
Comments	Carbon stars Classification is difficult for these stars, hence many C-type stars are also classified as N or R

Class S

Temperature	2000 K to 3000 K
Prominent spectral lines	TiO and ZrO bands
Examples	R And (S6ev), W Cet (S7,3e), X Aqr (S6,3e)
Comments	Long-period variable giants

Section 2

Practical Amateur Spectroscopy

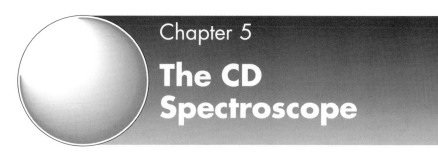

Chapter 5

The CD Spectroscope

David A. Randell

Introduction

This chapter describes how one can use CD or DVD compact disks to see and record spectra and their emission and absorption lines. To see these spectra, all you need is a CD, while to record these spectra, a 35 mm camera or equivalent is all that is required. Moreover, while some simple modifications and use of other equipment is suggested, none are absolutely necessary. So, arguably, the CD spectroscope has to be both the simplest and cheapest spectroscope one will ever come across. Using a CD and indeed using other common objects as inexpensive gratings is nothing new and has been reported before (Hecht, 1974; Gavin, 1997; Schwabacher, 1999), yet the CD's use as a simple inexpensive but *useful* spectroscope still remains relatively unknown.

Most will have held up a CD to some localized light source and noticed the rainbow-like colors that form as the disk is tilted back and forth. These originate in spectra resulting from the CD which is acting here as a *diffraction* grating. The CD incorporates a single spiral track that is pitted to encode digital information. The fine regular radial spacing between neighboring tracks in the aluminum layer forms the reflective grating which reflects the incident light, to form spectra. The number of radial tracks and the fine spacing between them is sufficient to resolve myriad emission and absorption lines to a surprising level of detail, even to the naked eye.

Figure 5.1 is an example of what can be achieved using this simple instrument. The solar spectrum shown here was captured on color print film using a 35 mm camera. Note that the color photographs reproduced here have *not* been digitally enhanced. The absorption lines are easily visible as dark line-arcs against the continuous colored backdrop, with some prominent ones easily visible in the original print labeled in Fig. 5.1a. The arc-like shape of the spectra themselves is a direct result of the circular nature of the CD-grating. While many more lines than shown here are visible in the original print, these are few compared to the mass of very fine lines that can be resolved when viewed directly.

CD (and DVD) Media as Diffraction Gratings

The typical compact disk (CD) is made from a small piece of plastic, coated with a thin reflective layer of aluminum, of some 120 mm in diameter and 1.2 mm in thickness. It carries a single spiral track that is pitted to encode digital information. The regular radial spacing between neighboring tracks (1.6 μm for CD media, and 0.74 μm for DVD) generates a set of closely and equally spaced lines which can function as a diffraction grating. As with many commercial gratings used in spectro-scopes, the CDs one buys are replication gratings, that is to say they are made from a master disk. CD and DVD media match commercially available gratings in terms of their effective number of rulings per unit length, making them well suited for use in simple spectroscopes. They also have the advantage of being minimal, if not zero cost, which contrasts with their research grade counterparts which are expensive, and not easily available (North, 1997).

Incident light falling on the grating is broken up into its individual wavelengths and spread out to form spectra of different orders as shown in Fig. 5.2.

Where *normal incidence* arises, i.e. where the incident light falls perpendicularly to the surface of the grating, the CD grating satisfies the *grating equation*: (i) $a \sin \theta = m\lambda$. However, where *oblique incidence* occurs, the general form of the equation for both *transmission* and *reflection* is used: (ii) $a(\sin \theta_m + \sin \theta_i) = m\lambda$. As our use of the CD

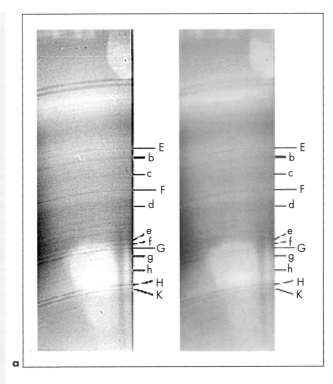

a

b

Figure 5.1 Solar spectra photographed using a CD-grating. Several prominent absorption lines (**a**) are visible in each photograph. The light vertical line in **b** and lobe-like patches in **a** and **b** are simply spurious reflections, and form no part of the main spectrum. The monochrome image (**a**, left) is produced from (**a**, right) and has been digitally enhanced to better show the absorption lines.

typically assumes a shallow grazing angle to the CD, it is the latter equation that we use.

where: θ_i = angle of incidence to the grating normal

θ_m = angle of transmission/reflection to the grating normal. $_+\theta_m$ means the angle of the deviated ray is on the same side as the incident ray and $-\theta_m$ on the opposite side.

a = grating ruling spacing

m = spectrum order

λ = wavelength of light.

The grating equation implies that *constructive* interference can occur for wavelengths corresponding to the orders 1, 2, 3,..., n. This means a grating produces several spectra, of which some will overlap. In the limiting case the angle of reflection becomes equal and opposite to the angle of incidence, and a direct (mirror like) image of the light source is produced. This is the zeroth-order case. Successive spectra orders (first, second, third, etc.) are increasingly deviated, dispersed and faint.

The grating equation also implies that the greater the number of rulings in the grating, the more the spectra generated will be deviated and dispersed, as depicted in Fig. 5.2. From this we would expect a larger deviation and dispersion for a DVD as opposed to a CD grating for a given order, and indeed this is the case.

The energy transmitted by the grating for the zeroth order makes up most of the net total energy

Figure 5.2. Spectra being formed by a reflective grating. A shallow grazing incident angle is assumed here. The incident ray is coming in from the right and forms angle θ_i to the grating normal and is diffracted (θ_m) into various orders: $m = 0$ (direct image), $m = -1$ (first order), $m = -2$ (second order), etc. The deviated angle θ_m to the grating normal is shown for the first-order spectrum.

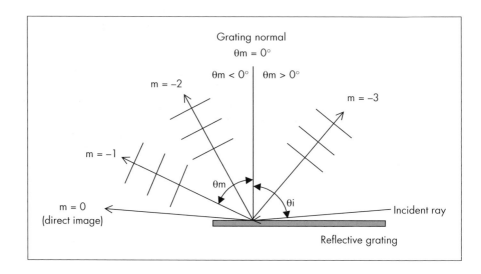

transmitted, the rest is distributed amongst the various orders. Generally, the higher the order, the less the energy carried, and the lower the brightness of its spectrum.

The CD Spectroscope in Use

Spectra are either continuous or discrete. The solar spectrum shown in Fig. 5.1 is an example of the former. The various wavelengths of light that make up these spectra form lines that together make up the colored continuum with which we are familiar. In addition bright or dark lines respectively known as *emission* and *absorption* lines, are often seen.

Spectra showing a discrete pattern of bright lines are known as *emission spectra*. Here the signature of the lines is used to infer what elements are making up the light source. A fluorescent light will produce such a spectrum. In contrast, a spectrum showing dark lines shows evidence of absorption of certain wavelengths of light transmitted by elements making up that source. Here the incoming light passes through a nonopaque body, such as a body of gas between the light source and (in our case) the grating used. The Sun and stars typically form such spectra.

You can use the CD to reveal absorption and emission spectra from numerous sources. Of these the solar spectrum is arguably the most interesting to see, though one needs to take very great care when using the Sun as the source of light otherwise irreversible damage to one's eyesight or even blindness can result.

Under no circumstances should you directly visually observe the zeroth or first order solar spectra. In addition, the CD spectrograph should not be used in conjunction with a telescope or binocular or other light-gathering instrument for unfiltered observation of solar spectra. Imaging solar spectra is covered separately (below) and you should not attempt it until you are absolutely clear what not to do and are certain that you have fully understood the potential risks and have taken steps to protect yourself against them.

To see spectra, it is easiest to start experimenting with artificial light sources, and in this respect distant streetlights at night make excellent targets. Using these you will see discrete emission spectra in the form of sets of bright colored arcs, where each individual line represents a particular element emitting light, while the whole functions as a signature identifying the nature of the type of light being used.

The general viewing method is as follows. First place the CD flat on the palm of your hand, with the silver side of the CD facing up and horizontal. Next you need to bear in mind two important points. The first is that in order to see spectra fanning out in a similar manner to that shown in Fig. 5.1, you need to place your eye close to the CD surface, at no more than 10 cm distance; and secondly, that you need to tilt the CD so that the grazing angle the incoming light makes to the CD's surface is very, very shallow – see Fig. 5.2. Aim for a grazing angle of ~3°.

Initially, it is best to first hold the CD vertically, then facing the light source, look at the surface of the CD and let your hand slowly assume a horizontal position. One needs to note the changing reflections while one does this. At some point the direct image of the light source will become visible. This is the zeroth-order spectrum. Now reversing the tilt direction, and passing through the zeroth-order point, a bright sliver of colored light (blue first then passing through yellow to red) will eventually appear and be seen to move from the perimeter of the CD to its center. This is the first-order spectrum. If using CD media, continue until you see a second (second-order) spectral streak form. Once you see that, maintain the same relative viewing angle to the CD surface, and making sure at all times you can see that sliver of light, try varying the distance from your eye to the CD surface. Do not rule out bringing your eye very close to the CD surface. At the optimal angle and distance, the spectral streak of light will suddenly appear to fan out to form around an apparent 60-degree arc; whereupon the spectrum and its absorption or emission lines will be immediately obvious. When you see the spectrum, it is worth slightly refining the tilt of the CD to make the grazing incidence angle as shallow as possible while continuing to keep the spectrum in sight. At some point shadows formed by the pitted tracks of the CD will be seem to run across the disk, which indicates the practical grazing angle limit has been reached. But by minimizing this angle, the visibility and the individual sharpness of the lines will be found to reach an optimum level.

To successively image first, second and third-order spectra you need to progressively steepen the viewing angle as shown in Fig. 5.2. For, example, to see the second-order spectrum using a CD one looks almost face on the disc, while for the third order you need to effectively bring the disk back towards you while maintaining the same angle the CD makes to the incident light, and now "look back" toward the disk perimeter.

It is relatively easy to prove to oneself that these lines originate in the spectra and not as artifacts arising from the CD itself, for as the CD is slowly tilted back and forth, the spectra and lines will be seen to move together, thus ruling out that possibility.

Solar Spectra

The *safest* way to view solar spectra is to use some other instrument to image the reflected light and *not* use direct vision. **You must never look at the Sun directly, or at its direct reflection (zeroth-order spectrum) off the CD or DVD disk.** Moreover, it is also strongly recommended that you do not attempt to directly look at **unfiltered** first-order solar spectra with the CD, as the spectra, though dispersed, will still be extremely bright. However, with very great care, one can see the solar spectra provided the above advice is rigorously followed.

Recording Spectra on Film

In the case of using a camera to record spectra, the same principle and viewing angle discussed above applies. For the solar spectrum shown in Fig. 5.1, a 50 mm lens was used coupled to a 35 mm camera. Using this set-up, the second-order spectrum generated by a CD, and first order using DVD, easily fills the 35 mm frame. You will need to bring the camera very close to the CD in order to achieve this. For example, to get the photograph shown in Fig. 5.1, the CD was actually touching the outer housing of the camera lens – see Fig. 5.3. Using different shorter/longer focal length

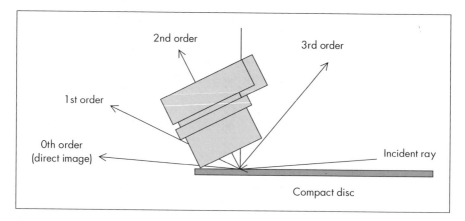

2nd order

3rd order

1st order

0th order
(direct image)

Incident ray

Compact disc

lenses will enable more or less of the spectra to be imaged; however the standard 50 mm lens coupled to a 35 mm camera was found to be the best lens/camera combination in practice.

In each case the camera lens is focused on *infinity*. This means one is not restricted to using cameras with lenses that focus, or indeed cameras with a through-the-lens imaging capability. However, the latter, while not absolutely necessary, is highly desirable, especially considering the following. If the camera has a through-the-lens imaging facility, it is better to center the image in the viewfinder while using the lens *stopped down* to the correct aperture for the intended exposure. Most cameras assume full aperture focusing by default to aid focusing and composition, but you need to stop down to the correct aperture while composing the shot if at all possible. If you do not do this, while you may well see the spectrum filling the viewfinder at full aperture, you may find that only a small part of the image will form on the film from vignetting; and in some cases will simply not record anything. The tolerance on the camera lens–CD alignment angle is very tight, and only when visually composing the shot using the same aperture for the correct exposure can one be assured of success.

Figure 5.3. Using a 35 mm camera to record second-order spectra. Notice the very close position of the camera lens housing to the reflection grating, and its position relative to the perimeter of the disk.

Modifications

The CD spectroscope as described does not require a slit, nor a collimator. The surprise is that fine absorption lines can be clearly identified and resolved using both an extended light source, and without using a slit. Its ability to resolve lines in spectra relies in part on the relatively small angular extent of the light source and the very narrow grazing angle of the incident light. However, the use of a slit common to most spectroscopes can easily be introduced if required. Using an optional slit will serve to clean up spectra, but is not necessary in order to resolve numerous absorption and emission lines.

For CD and DVD disks, a thin horizontally positioned slit can be easily made from two pieces of thin cut card, inserted into the clear top of a standard CD case. By using the two separate pieces of card, one can easily vary the width of the gap between them as required. With the CD clipped into the case face up, it makes for a very compact and portable instrument. By adjusting the angle of the lid relative to the case, and the width of the slit, the brightness and the clarity of the resolved lines will vary as a direct result. Using the slit will significantly cut down the amount of incident light falling on the CD and greatly improve the detection and resolution of lines visible in first-order continuous spectra.

A second refinement is to completely seal the CD in a card box. You will need to work out the respective angles required to position the viewing holes, either empirically or analytically using the grating equations introduced earlier. By constructing this box you can make it impossible to image the zeroth-order image (which can be useful if imaging very bright sources) while making it much easier to see different order spectra, as much of the alignment process will have already be done. An excellent example of this type of CD spectroscope has been designed by Alan Schwabacher, where he provides complete downloadable instructions and figures that can be copied on to card, cut out, and assembled (Schwabacher, 1999). Obviously, if using DVD media, the various precomputed angles applicable for a CD grating will need to be recalculated.

The use of simple colored filters common to both photography and visual planetary astronomy may help in the identification of some spectral lines. Here the filter intercepts the light stemming from the grating before it reaches the eye or film in the camera, and hence only passes specific wavelengths. This can be useful where overlap of orders makes visual identification of some lines difficult if only using the colored backdrop of continuous spectra as a guide. A clear example of this can be seen in Fig. 5.1a. Simple extrapolation from positively identified lines shows that the position of line D for the second-order spectrum shown, coincides with one of the prominent pair of absorption lines shown at the top of the photograph. However, the line pair matches the predicted positions of H and K resulting from overlapping third-order spectra; and in fact appear to be their true origin. Another simple and related technique (if you have access to a computer, scanner and image processing software) is to scan in and digitize the image, and then split the image into their RGB components (or channels). For example, using this technique on the original print, the blue channel immediately revealed the existence of an otherwise invisible third-order line neighboring H and K, but which mirrored the relative position of the innermost prominent line clearly visible in the photograph. This confirmed the hypothesis that the two lines shown in the photograph originate from the overlapping third-order spectrum.

Finally, one can use the CD spectroscope to explore the emulsion sensitivity of films, and check the passbands of light for specific interference and light pollution filters. In all cases, spectra generated give an unambiguous and measurable way to analyze and compare the results.

Orders of Spectra, Overlapping and Resolution

Using the CD grating, first and second-order spectra are reasonably separated, but second, third and higher-order spectra perceptibly overlap. With the CD it is just possible to see lines originating from fifth-order spectra. In comparison, using the DVD grating, first and second-order spectra, and barely part of the third order will be visible. However, first and second-order spectra using DVD will have a much larger dispersion than those generated by a CD grating. A comparison of angular dispersion and overlapping of orders for CD and DVD media is shown in Fig. 5.4.

Here the shallow grazing incidence angle ($\theta_1 = 87°$) now works to one's advantage, since the theoretical range of viewing angles effectively doubles from ~90°

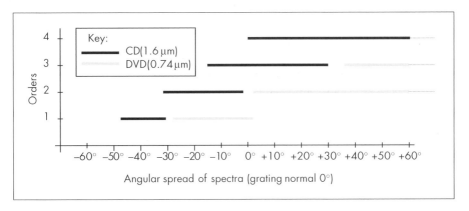

Figure 5.4.

Overlapping orders of spectra generated by CD and DVD gratings $(\theta_i = 87°)$. Each spectrum covers 3934 Å (line K) through to 7594–7621 Å (line A).

(for normal incidence) to ~180° (shallow grazing angle); noting that for normal incidence the imaging of higher-order spectra is very restricted.

The resolution and the detection of individual lines will depend on the type of spectra, the grating used, and the order imaged. And when used by the direct vision method, it will also depend on the acuity of the observer's eyesight. With good eyesight, it should be relatively easy to split the Na doublet (line D) at 6 Å separation, and all three components of the Mg triplet (line b) at 11 Å and 5 Å separations for the main and secondary line pairs. This assumes using at least second-order spectra with the CD and first-order spectra using DVD media. Theoretically, greater resolution of the lines is possible using higher-order spectra but in practice overlapping of the orders will generally degrade their clarity.

Apart from the obvious advantage of using film to make a permanent record of spectra, film enables one to record faint or near invisible (in the visual) spectra by the simple expedient of using long exposures. Photographic resolution of spectral lines for comparable orders will generally not equal that seen by the naked eye and will typically require greater dispersion to be resolved on film. However, overlapping orders will reduce image contrast and can make the lines more difficult to positively identify. Furthermore, while film emulsion may gain by revealing the presence of lines barely visible to the naked eye owing to spectral sensitivity differences, it can equally fail in other respects. For example, when using a CD-grating, the prominent lines H and K shown in the photograph in Fig. 5.1a are not easily visible (at least to this author); while line D, although clearly identifiable when viewed directly, cannot be similarly identified in the same photograph.

However, these limitations aside, it is worth pointing out that when closely examined, the original print shown in Fig. 5.1a shows the main pair of lines of the Mg triplet (at 11 Å) just resolvable, while at least 47 separate lines in various groups can be positively identified. This gave an estimated maximum and average combined grating and photographic resolving power of ~1000 and ~630. It is also worth noting that the set of photographs reproduced here was the very first attempt by this author to record spectra using a CD grating. Moreover, no attempt had been made to select either a very fine grain film or one singled out for good linear spectral response. In this respect a CCD camera with its well-known excellent linear response would be expected to have a definite edge over film.

Some Suggested Targets

For *emission spectra*, there are many artificial sources that can be readily used. Imaging street lights, for example, transforms their light into a distinctive set of very colorful and discrete arcs. Try incandescent and fluorescent lamps, street lights (mercury, low-pressure sodium, and high-pressure sodium) and neon signs. Even a simple candle flame will provide enough light to reveal its bright yellow sodium line. Where the light source is large and diffuse, the lines will become fuzzy and indistinct, and often cannot be resolved. Aim for a source of light that does not subtend more than 0.5°, which is the angular size of the solar disk, and which, as has been shown here, is small enough to reveal many fine low-contrast absorption lines. Where the source of light subtends an angle greater than this, a mask can be employed, or alternatively one can simply introduce a slit if the source is bright enough to see or record the spectra.

Imaging *absorption lines* in solar spectra has already been covered. However two points relating to solar spectra are worthy of note. Comparing solar spectra with the Sun high overhead, and low down near the horizon will show marked differences. When low on the horizon, sunlight passes through a denser layer of the atmosphere, and the increase of the absorption of the incident light by water vapor will now show up as a very prominent group of thick absorption lines in the deep red end of the spectrum. Secondly, having a CD to hand during a total eclipse of the Sun may be worth trying out to see if one can record solar emission lines in flash spectra

Conclusions

CD and DVD media offer an inexpensive and simple way to obtain reflection gratings and make a very simple spectroscope. Using nothing more than a simple camera, complex structure in spectra can be easily be recorded, measured, and analyzed. Apart from its novelty value, the CD spectroscope may be of interest to physics educators introducing science students to spectroscopy, particularly where simplicity of set-up and minimal cost is of paramount importance. The modifications and experiments mentioned here fall well within the capabilities of most people, and assume very little practical or dexterous skill, while the additional equipment mentioned is relatively inexpensive and typically easily available.

Bibliography

Gavin, M (1997) *Amateur Spectroscopy*, Journal of BAA, vol. 108, #3, June 1998; October 1997 Presidential Address. See also Maurice Gavin's introduction to spectroscopy: *Spectroscopy – A Practical Beginners Guide*: www.astroman. fsnet.co.uk/begin.htm

Hecht, E (1974) *Optics*, 3rd edn. Addison-Wesley, p. 430.

Nelkon, M and Parker, P (1970) *Advanced Level Physics*, 3rd edn. Heinemann Educational Books.

North, G (1997) *Advanced Amateur Astronomy*, 2nd edn. Cambridge University Press, p. 348.

Schwabacher, Alan (1999) *Mini Spectroscopes*. University of Wisconsin-Milwaukee. See: www.uwm.edu/~aawschwab/ specweb.htm.

Sidgwick, JB (1980) *Amateur Astronomer's Handbook*, ed. James Muirden, 4th edn. Enslow Pub.

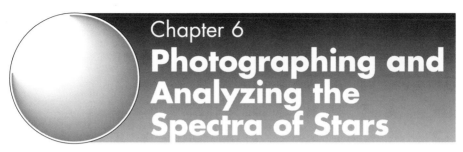

Chapter 6
Photographing and Analyzing the Spectra of Stars

Jack Martin

Equipment

Splitting starlight and photographing the resulting spectrum of a star is not as difficult as one may think. My set-up consists of a 0.30 m f/5.3 Dobsonian telescope, grating, grating mount, low-profile helical focuser, tele-extender (variable focus eyepiece projection tube), T-ring, 35 mm SLR camera body, a suitable detector, in this case 35 mm black and white film, a long air cable release, electronic timer, LED light, notepad, pen, table and chair.

The telescope, a Newtonian reflector, is attached to a Dobsonian mount – see Fig. 6.1 – which moves in altitude and azimuth only. As seen in Fig. 6.2, the telescope has been modified by moving the focusing rack mounting plate forward, in order to achieve a focused image at the prime focus. The reader should note that this type of photography cannot normally be done with a standard Newtonian reflector, and that the Schmidt–Cassegrain design is more suitable, because it has a deeper field of focus. It is also important for the reader to note that results were obtained using a 0.30 m primary mirror. A smaller aperture primary mirror may not give as good a result, due to its inferior light gathering power and resolution.

The grating is a Rainbow Optics slitless transmission grating, which only works on stars and not on extended objects. The grating mount is simply a piece of internally threaded aluminum tube which the grating screws into, which in turn is housed inside a tele-

Figure 6.1.
0.30 m f/5.3
Dobsonian telescope
and with camera/
grating assembly.

extender and is locked securely in position by a thumbscrew. The tele-extender is especially well suited to this type of photography because it allows the camera/drawtube assembly to be rotated whilst the grating is in the locked position. The tele-extender is of the variable focus type, so when the drawtube is pulled back to enlarge the image, it will go out of focus. The only way to focus the enlarged image is to move the focusing rack mounting plate further forward. But because of the way a Newtonian reflector is designed, there are limits as to how far forward the focusing rack mounting plate can be moved. It is important to bear in mind that the more the image is enlarged the fainter it becomes. I have found that a smaller brighter image is better than a fainter larger image. So, for this set-up the drawtube focus is always set at the minimum image magnification. Should the reader decide to make such a modification, then it is simply a matter of trial and error

Figure 6.2. Modified focusing rack mounting plate assembly.

Figure 6.3. Olympus OM1-N 35 mm SLR camera body, T-ring, grating housing, Rainbow Optics grating, Tele-extender.

to achieve the optimum image size for their particular Newtonian reflector. A T-ring is needed to attach the tele-extender to the 35 mm SLR camera body, see Fig. 6.3, and a low-profile helical focuser was also fitted in order to achieve the depth of focus as seen in Fig. 6.4.

The camera body is an Olympus OM-1 N, which is particularly well suited for astrophotography because it is a lightweight mechanical camera, that does not use a battery, except for the light meter which is not used for astrophotography, so the shutter can remain open for as long as necessary. Also, the focusing screen is interchangeable, and Olympus makes a no. 8 focusing screen especially for astrophotography.

The detector is 35 mm black and white film, which is discussed in more detail under the next subheading "Types of Black and White Film".

A long air cable release is useful since there is no need to be next to the telescope whilst the shutter is open. This means you can sit down comfortably, while recording the star's spectrum. Only an LED type light should be used when writing, as this particular light does not interfere with night-adapted vision. A notepad and pen are a must for writing details of timings and stars being photographed. An electronic timer is important for accurate measurements of drift times across the field of view. I recommend the Jessops timer for this purpose. A small table and chair make life easier when there is a need to put things down in one place, or when sitting down to write up details.

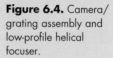

Figure 6.4. Camera/ grating assembly and low-profile helical focuser.

Types of Black and White Film

Where black and white photography is concerned, there are five different film emulsions:

- Unsensitized – sensitive to ultraviolet, violet and blue light;
- orthochromatic – sensitive to blue and green light;
- panchromatic – sensitive to blue green and red light, i.e. the entire range of visible light, approximately 4000–7000 Å;
- hyperpanchromatic – panchromatically sensitized with increased sensitization to red light;
- infrared – sensitive to the nonvisible infrared radiation.

Panchromatic films are best suited to, and the most easily available films for, stellar spectra photography. The six factors, which affect the result, are: film speed ISO, spectral sensitivity of the film, color and magnitude of the star, phase of the Moon, and aperture of the telescope primary mirror. The general rule of thumb is to use slow films for bright stars, and fast films for fainter stars. The reader must experiment with different speed films to see what works best for their particular set-up. The spectral sensitivity of different panchromatic films can vary. Figure 6.5 shows the spectrum of ζ U Ma A+B taken on Fuji Neopan ISO 1600/33° and Ilford Delta ISO 3200/36° films, the second of which has

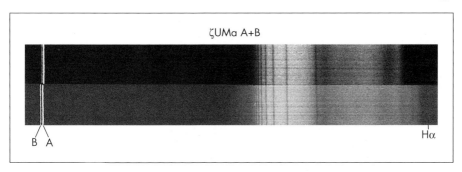

Figure 6.5.

Spectrograms of ζ U Ma A+B spectral types A2 and A1 taken on Fuji Neopan ISO 1600/33° and Ilford Delta ISO 3200/36° showing the Hα line and the A and B components of this spectroscopic binary.

Figure 6.6.

Spectrograms of α C Ma spectral type A1 temperature 9400 K taken on Ilford FP4 Plus ISO 125/22° and Ilford SFX ISO 200/24° showing the Hα and O II (telluric oxygen) lines.

a wider spectral sensitivity both at the blue and the red end of the spectrum and shows the Hα 6563 Å line at the red end. Figure 6.6 shows the spectrum of α C Ma taken on Ilford FP4 Plus ISO 125/22° and Ilford SFX ISO 200/24° films, the second of which has extended red sensitivity up to 7400 Å and shows the Hα 6563 Å and O II 6870 Å (telluric oxygen from the Earth's atmosphere) lines. Film manufacturers technical fact sheets usually show a spectral response curve known as a Wedge spectrogram, which indicates the relative sensitivity of an emulsion at different wavelengths through the spectrum. The cooler stars do not photograph as well as the hotter stars, because the panchromatic film is less sensitive to the cooler star's colors, an all-important factor, which is ultimately determined by the star's temperature. This type of photography is best done when the Moon is out of the way, so moonrise and moonset tables are needed for this purpose. Also, sunrise and sunset tables should be used to determine when full darkness begins and ends, before starting a photographic session. A larger aperture mirror will give a brighter image than a mirror of smaller aperture.

Color films are not recommended for stellar spectra photography, because the tricolor emulsion has the effect of masking the spectral lines, and the response at

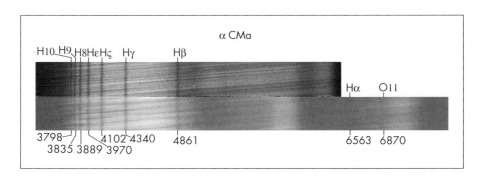

the blue end is poor. As a result very few spectral lines are visible.

Method of Photography

Point the set-up at a bright star, loosen the helical focuser thumbscrew, then rotate the camera/grating assembly clockwise or anticlockwise until a sharp focus is achieved, then lock the helical focuser thumbscrew, see Fig. 6.4. Now move the set-up across to the star's first-order (brightest) spectrum. Loosen the two tele-extender drawtube thumbscrews and rotate the drawtube/camera assembly until both the star and its spectrum is seen running parallel and next to the short edge of the camera viewfinder, now lock the two thumbscrews. Then keep rotating the camera/grating assembly following only the star by eye, until the star trails across the field of view parallel and next to the long edge of the camera viewfinder. In practice this is done by checking the star's position and adjusting the camera/grating assembly clockwise or anticlockwise accordingly, by loosening the thumbscrew, which attaches the tele-extender to the helical focuser, see Fig. 6.4. Finally, repeat the focusing procedure on the same star. Once focused, there is no need to repeat the procedure for other stars. To record the star's spectrum allow the star and its first-order spectrum to drift across the field of view, until the spectrum is widened sufficiently enough to show the detail of the spectral lines as seen in Fig. 6.7. This can vary from 15 seconds to 5 minutes, using diurnal motion, and depending on the proximity of the star to α U Mi, since all stars at this latitude appear to revolve around α U Mi, the Pole star, as seen in Fig. 6.8.

Figure 6.7.
Spectrograms of α Lyr spectral type A0 temperature 9500 K taken on Ilford FP4 Plus ISO 125/22° showing the widening effect from 1 to 15 seconds.

Figure 6.8. The movement of stars around α U Mi, the Pole star.

Development and Printing of Film

All developing and printing is done in my darkroom. The development process is exactly the same for all the different makes of black and white films used. The only variant is the development time for each film. Paterson Aculux 2 very fine grain developer is used to process all films, adhering strictly to the manufacturer's recommended development times and temperatures. Push processing these types of pictures is not recommended and does not work for stellar spectra photography. To stop the film development process, Ilford Ilfostop odorless stop bath is recommended. For fixing the film use Ilford Hypam rapid fixer. The film is then washed for the specified time and left in Ilfotol wetting agent for 1 minute, then dried with a film squeegee, hung on film clips, and left to dry in a warm dust-free cupboard. Once dry, use only lint-free cotton gloves to handle the film, and cut into strips of six frames per strip. The negatives are placed on a light box; using a 10 × loupe, the best negatives are then selected for printing.

The width of the spectral image on the negative varies, depending on the spectral sensitivity of the film, which is

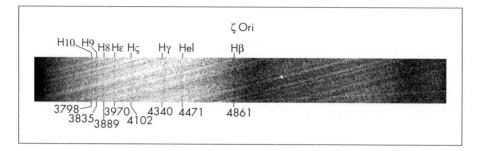

then enlarged to the same magnification as the spectral lines of a guide print of a star of the same spectral type, as seen when superimposed on the negative image, to a dispersion of 17.8 cm, the length of the photographic paper used. Jessops Grade 3 gloss resin coated monochrome photographic paper size 12.7 × 17.8 cm is recommended, and developed in Jessops Photochem Econoprint paper developer. The stop bath and fixer used are the same as mentioned above. Further examples of my spectrograms are shown in Figs 6.9–6.16.

Figure 6.9.
Spectrogram of ζ Ori spectral type O temperature 30 000 K taken on Fuji Neopan ISO 1600/33°.

Processing Black and White Slides from Black and White Negatives

Black and white slides of stellar spectra are very useful for presentation purposes. Agfa Scala ISO 200/24° is the only 35 mm black and white slide film available, but only works on bright stars. It cannot be processed in-house. Only Joe's Basement in London can process this film.

Figure 6.10.
Spectrogram of γ Peg spectral type B2 temperature 22 000 K taken on Fuji Neopan ISO 1600/33°.

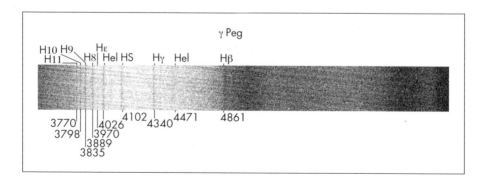

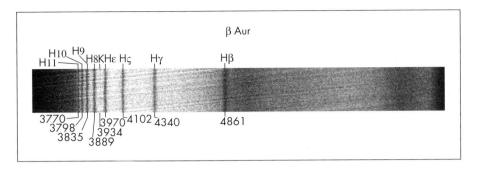

Figure 6.11. Spectrogram of β Aur spectral type A2 temperature 9000 K taken on Fuji Neopan ISO 1600/33°.

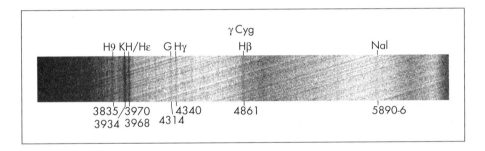

Figure 6.12. Spectrogram of γ Cyg spectral type F8 temperature 6100 K taken on Fuji Neopan ISO 1600/33°.

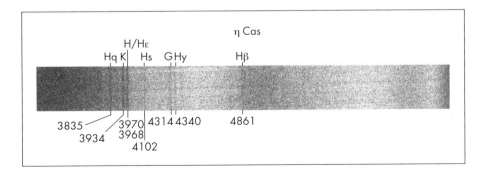

Figure 6.13. Spectrogram of η Cas spectral type G0 temperature 5900 K taken on Fuji Neopan ISO 1600/33°.

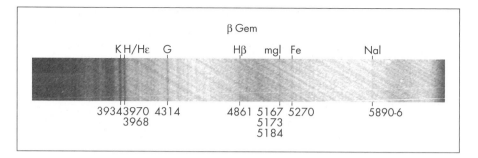

However, in the process described below a black and white positive can be made from any make or speed of black and white film. To do this the following equipment is needed: a light box, tripod, 35 mm SLR camera body, T-ring, slide copier, long air cable release, electronic timer, notepad, pen, table and chair.

Figure 6.14.
Spectrogram of β Gem spectral type K0 temperature 5100 K taken on Ilford HP5 Plus ISO 400/27°.

The negative is inserted into the slide holder of the slide copier. The image is then lined up, centered and magnified as necessary through the camera viewfinder. Once set up, open the shutter for 25 seconds, and repeat the same time exposure for the same or any different negative. The film used is Kodak Eastman fine grain release positive film, and processed in Jessops Photochem Econoprint photographic paper developer. Do not use a film developer, as it does not work on this type of film. The film is then stopped, fixed, washed and left in wetting agent using the same chemicals as mentioned in the "Developing and Printing of Film" subheading, and dried in the same way as any other film. Unlike other films, it can be safely exposed under a red safelight. The positives are then cut and mounted using lint-free cotton gloves.

In the darkroom, it is important to follow the manufacturer's processing recommendations and health and safety procedures at all times. Always use fresh solutions for processing. Never use the same set of chemicals for films and papers.

Figure 6.15.
Spectrogram of α Ori spectral type M2 temperature 3400 K taken on Ilford HP5 Plus ISO 400/27°.

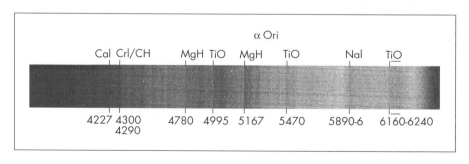

Figure 6.16. This composite of spectra of M 45, The Pleiades, consisting of stars of spectral type B was taken on Ilford Delta 3200/36°.

Analysis of Spectrograms

There are two methods of analysis that can be used. Both rely on two known lines, and the fact that the relationship between the distance and the wavelength of the spectral lines are linear when a diffraction grating is used as a disperser. This means that simple linear equations are used to calculate the wavelengths of the lines of the elements present.

Method 1

Three simple linear equations are used as follows:

1. $a = w_z/d$
2. $w_z = d \times a$
3. $w_x = w_y w_z$

where

a = average dispersion;
d = distance difference between the same two lines;
w_x = wavelength of the line;
w_y = wavelength of the base line;
w_z = wavelength difference between two known lines.

To find the wavelength difference w_z, simply subtract the wavelengths of the two known lines. The distance difference d is determined by measuring the distance between the two known lines with a ruler. The answer gives the average dispersion in angstroms per millimeter. Use the two furthest lines at the blue and the red end of the spectrogram to get the most accurate figure for the average dispersion. The use of either the + or – sign will depend on which end of the spectrogram the base line is taken from. The answer gives the wavelength in angstrom units (Å) of the element present. A steel rule graduated to 0.5 mm, a magnifier and calculator is needed for this exercise. Two publications for verifying the results obtained are recommended:

1. W.C. Seitter, *Atlas of Objective Prism Spectra*. Ferd Dummlers Verlag, Bonn, 1970.
2. C.E. Moore, *A Multiplet Table of Astrophysical Interest*. NBS Technical Note 36, United States Department of Commerce, revised edn, 1972.

Example   measurements from the spectrogram of α Aqu Fig. 6.17 using Na I and H II as the two known lines:

$a = w_z/d$
$w_z = $ Na I – H II (Å)
$w_z = 5890 - 3770 = 2120$ Å
$d = $ Na I to H II (mm) $= 123$ mm
$a = 2120/123 = 17.23$ Å m^{-3}
The base line w_y is Na I $= 5890$ Å
The calculations are given in Table 6.1.

Table 6.1. Calculations for α Aqu

Line	Calculated Value w z (Å)	Calculated Value w x (Å)	Accepted Value (Å)	Error (Å)
$H\beta$	$60 \times 17.23 = 1033.8$	$5890 - 1033.8 = 4856.2$	4861	−4.8
$H\gamma$	$90 \times 17.23 = 1550.7$	$5890 - 1550.7 = 4339.3$	4340	−0.7
$H\delta$	$104 \times 17.23 = 1791.9$	$5890 - 1791.9 = 4098.1$	4102	−3.9
$H\varepsilon$	$111.5 \times 17.23 = 1921.1$	$5890 - 1921.1 = 3968.9$	3970	−1.1
K	$113.5 \times 17.23 = 1955.6$	$5890 - 1955.6 = 3934.4$	3934	+0.4
H_8	$116 \times 17.23 = 1998.7$	$5890 - 1998.7 = 3891.3$	3889	+2.3
H_9	$119.5 \times 17.23 = 2059.0$	$5890 - 2059.0 = 3831.0$	3835	−4.0
H_{10}	$121.5 \times 17.23 = 2093.4$	$5890 - 2093.4 = 3796.6$	3798	−1.4

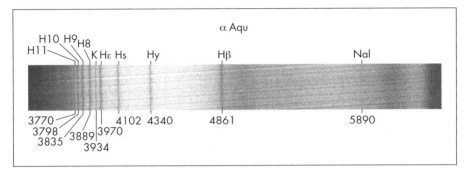

Figure 6.17.
Spectrogram of α Aqu
spectral type A7
temperature 8100 K
taken on Ilford FP4 Plus
ISO 125/22°.

Method 2

This involves the use of a computer and spectrographic analysis software available free over the Internet from Valerie Desnoux, the author of *Visual Specs*. Although intended for the analysis of CCD spectral images, it will also read digitized photographic spectral images converted from a BMP file to PIC format using Christian Buil's IRIS program. Martin Peston was the first person to use these two sets of software to convert my photographic spectrograms to graphs, and write the following set of instructions:

Creating a Graph from a Spectrogram

1. Scan in a spectrogram. The resolution does not need to be high: 100 dpi is acceptable.
2. Save the file as a BMP file.
3. Start up the IRIS program.
4. Load the BMP file into the program by selecting FILE then LOAD.
5. Save the file as a PIC by selecting FILE then SAVE then select the "save as type" option to select the PIC option.

6. Start up the *Visual Specs* program.

7. Open the PIC file that was saved by the IRIS program.

8. Click on the "Reference Binning" button denoted by ⬚ (see below).

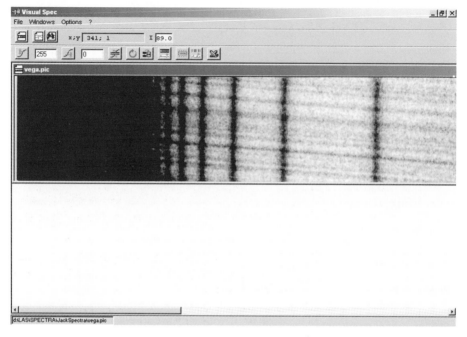

This will bring up another window depicting a graph.

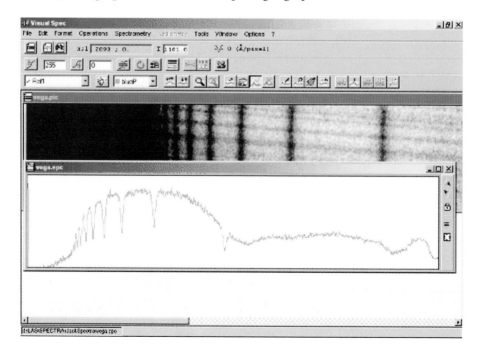

9. Now, select "Options" from the top menu bar, and then select the "Preferences" option.

10. Select "References" tab option.

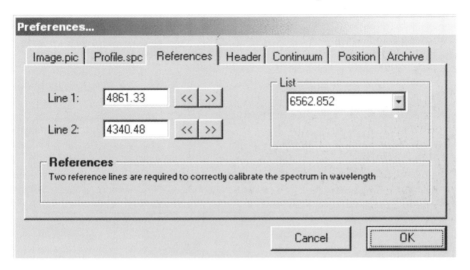

11. Enter the wavelengths of two known lines of the spectrogram in the Line 1 and Line 2 fields. Now calibrate the graph.

12. Select the "Spectrometry" option from the top menu. Then select the calibration option.

13. When the mouse cursor is moved over the graph a red line will also be seen moving along the graph.

14. On the graph take the mouse and make a rectangular dotted box around the first line, Line 1. This is done by taking the mouse cursor slightly to the left of the line, and with the left mouse button down move slightly to the right of the line. This line will have the value that was used in the preferences option.

15. Right click the mouse within the dotted box and there will be an option denoting "Line 1" or "Line 2". Highlight "Line 1" by selecting it.

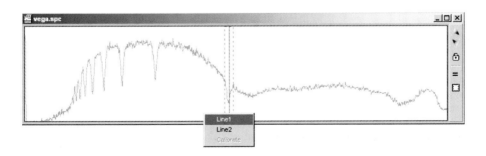

16. Select a second line that has been referenced and follow steps 14 and 15 but for "Line 2".

17. Right click the mouse button and select the "Calibrate" option.

18. Select the ▢ button which is on the far left of the graph to get the bottom x-axes to display values in angstroms.

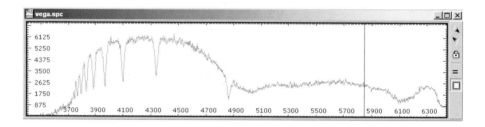

19. Double clicking on the graph brings up the following window.

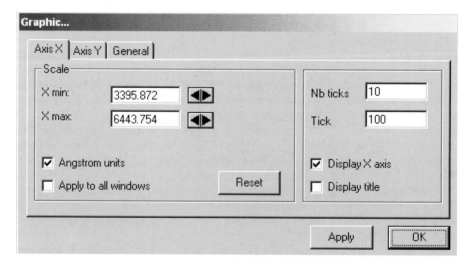

With these options both axes can be formatted with reasonable values as well as adding a title to the graph.

20. When the mouse cursor is moved over a line, the wavelength will be shown in the top window on one of the Visual Specs panels near the menu buttons.

21. The line spectra can now be printed by exporting to a BMP file and then printed from Word. This is

done by selecting "FILE" then the "Export to BMP" option. The final result is shown below:

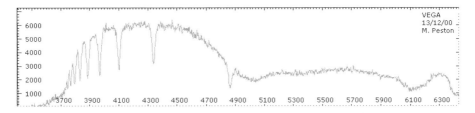

22. Use the down arrow on the right panel of the graph window to reduce the height of the graph.

23. To get a reference graph on to the same graph select "Tools" from the top menu then "Library". In the reference list select the spectral type which matches that of the star being analyzed and drag the spectral type list item over on to the graph.

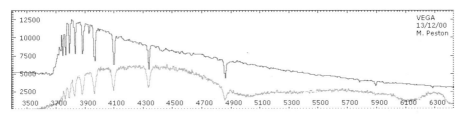

Conclusion

Once the graph has been calibrated there are many more analysis features available, such as identification and labeling of chemical elements, determination of the Plank temperature of a star, Doppler shift, speed of expansion, etc. The reader must study the reference manual for these and additional features.

Instructions courtesy of Martin Peston.

Further examples of my spectrograms converted to graphs by Neil Martin are shown in Figs 6.18 to 6.22.

The Astronomical Value of these Spectra

The spectra of stars are their own unique signatures. They are a valuable guide for amateur astronomers who wish to pursue this subject, and give a basic insight into

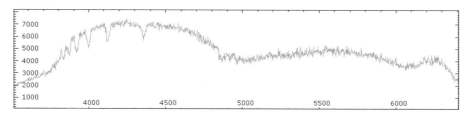

Figure 6.18. Spectrogram of α Leo spectral type B7 temperature 12 000 K taken on Ilford HP5 Plus ISO 400/27°.

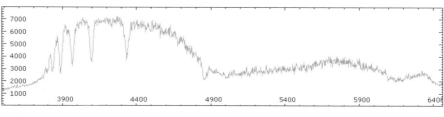

Figure 6.19. Spectrogram of β U Ma spectral type A1 temperature 9000 K taken on Fuji Neopan ISO 1600/33°.

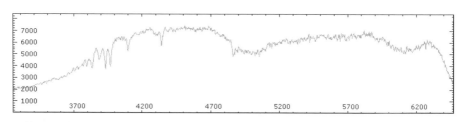

Figure 6.20. Spectrogram of α C Mi spectral type F5 temperature 6400 K taken on Ilford HP5 Plus ISO 400/27°.

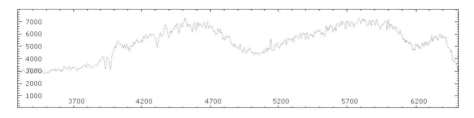

Figure 6.21.

Spectrogram of η Peg spectral type G2 temperature 5800 K taken on Fuji Neopan ISO 1600/33°.

the makeup of stars that is easy to read and understand. There use as educational aids should not be under-estimated. They are especially useful for teaching A-level physics and undergraduate astrophysics courses. They are also useful at astronomy exhibitions for amateurs or members of the public who are interested in astronomical spectroscopy. They can be used by the amateur for classification and identification purposes, using the two methods described under the subheading "Analysis of Spectrograms". There is also a need to provide basic information from a theoretical and practical viewpoint about astronomical spectroscopy for beginners and amateurs that is not available at this level in other books. Astronomical spectroscopy, traditionally the domain of professional astronomers, is an area of astronomy that very few amateurs would get involved in, because it was thought to be too difficult to do. This chapter and book shows that this is not the case. Useful results can be obtained with relatively modest equipment. I hope that further project work will be undertaken from the ideas in this chapter and book.

Figure 6.22.

Spectrogram of α Tau spectral type K5 temperature 4370 K taken on Ilford HP5 Plus ISO 400/27°.

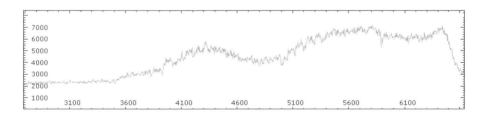

Chapter 7
Two Spectrographs for Small Telescopes

Nick Glumac

Introduction and Motivation

During the past 15 years or so, the appearance of new technology in the commercial marketplace has opened up an array of opportunities for those with access to small telescopes to explore the sky in new ways, educate others about astronomy, and participate in astronomical research. Perhaps the most significant of these new opportunities is in the area of spectroscopy. Spectroscopy has long been the primary tool in astronomical research. Many, if not most, of the major astronomical discoveries have been based upon spectroscopic observations. These discoveries include temperatures and compositions of stars, planetary atmospheres, nebulae, and comets, as well as relative galaxy and stellar velocities derived from red shifts. Despite the usefulness of spectroscopy as an astronomy tool, until recently, little attention has been given to making spectroscopic observations using small telescopes such as those used by amateur astronomers, small colleges, and high schools. The primary reason for this lack of attention is the difficulty in obtaining quality spectra of a wide enough range of objects using film due to the low sensitivity of photographic emulsions. While spectra of bright objects such as Jupiter and Venus can be obtained with moderate resolution on film with an 8-inch telescope, even low-dispersion spectra of objects dimmer than magnitude 5 are difficult. Thus, the lack of a suitably diverse set of targets on which to make observations, plus the difficulties of processing film and extracting quantitative data, has made spectroscopy a less attractive pursuit for the small telescope owner.

The potential for spectroscopic measurements exploded as CCD cameras became available. CCDs boast quantum efficiencies in the neighborhood of 20 times larger than photographic film, and their signal levels typically vary linearly with photon count, making processing much simpler. With a 20-fold increase in sensitivity, an 8-inch telescope with a CCD can, in principle, produce the same quality spectra as a one-meter telescope operating with film. Thus, almost overnight, the small telescope owner was given the opportunity to make the same scientific observations that were previously only achievable through the use of large observatory telescopes. Many of the key discoveries made using spectroscopy at the major observatories in the 1960s and 1970s could easily be replicated by a small telescope owner with a simple spectrograph and CCD camera in the early 1990s. In my own work, I have used an 8-inch telescope to repeat the observations of Spinrad and Trafton (1963), in which they discovered a new hydrogen line in the atmosphere of Jupiter, and those of Owen (1971), who discovered a new band of ammonia in Jupiter's atmosphere in 1969. Many new research avenues were also opened to those with small telescopes. Low-dispersion spectra of stellar objects out to about magnitude 12 can be used to analyze stars, identify novae and supernovae, and monitor variable stars with more than just an intensity versus time curve. In addition, the spectroscopy of comets remains an active area of university research, and many comets each year are bright enough for low and medium resolution spectroscopy with small telescopes.

The opportunities for education have also increased in the area of spectroscopy. High schools, colleges, and universities frequently have a telescope in the 10-inch class for astronomy labs in which basic astronomical observations are demonstrated. However, even the most fundamental spectroscopic observations (stellar classifications, spectral characteristics of different astrophysical objects, etc.), on which much of the astronomy education is based, are simply too difficult to perform with film spectroscopy. With a CCD camera and a simple spectrograph, these subjects are now readily demonstrated, and the development of advanced astronomy labs with simple research projects is now possible.

While the opportunities in spectroscopy may be clear, the implementation of these opportunities is hindered by the lack of availability of appropriate spectrographs for small telescopes. As of the mid-1990s, only one major manufacturer was producing a spectrograph for small telescopes. Since then, two other manufacturers have emerged, producing lighter and simpler spectrographs for use with a variety of CCD cameras. Still the cost of these units (typically between US $2100 and US $4000) remains high for many budget-conscious amateurs and educational programs, and the selection of three models is quite limited. For example, resolving individual spectral lines in the absorption features of planetary atmospheres requires resolutions of better than 0.5 Å, which is at least a factor of two lower than the capabilities of any of the three commercially available spectrographs. While scientific spectrographs are available in a wide variety of focal lengths and f numbers, few are lightweight enough to be adapted to small telescopes, and most are far more expensive than US $4000 when all the appropriate accessories are added.

My work in the area of astronomical spectroscopy has been to develop simple spectrograph designs that are usable with small telescopes for research and education. A major objective of my efforts has been to generate instrument designs that can be constructed using conventional fabrication techniques for several hundred dollars, which, based upon my conversations with other amateurs and educators, seems to be a reasonable cost for an in-house project. I have focused on both educational and research aspects in my designs. For those willing to undertake research projects, a fast, small spectrograph capable of taking low to moderate resolution spectra of dim targets is required, since most brighter objects (e.g. brighter than magnitude 6 or so) have already been exhaustively studied. For the educators who wish to demonstrate key features in the spectra of bright objects (e.g. the absorption lines in planetary atmospheres), a high-dispersion spectrograph that can be readily coupled to a small telescope is required. Since it is unlikely that a single instrument can achieve both objectives, I have split my efforts among two designs: a compact, research spectrograph, and a larger, high-resolution spectrograph. In both cases, I was able to design and build spectrographs that performed similarly to or better than commercial units, and yet could be produced for much lower costs using standard fabrication techniques. This chapter discusses the design, construction, testing, and performance of each of these two spectrographs.

Design Considerations for the Spectrographs

There are several common challenges involved in the design of inexpensive low and high dispersion spectrographs, as well as challenges that are specific to each type. Both spectrographs require high-quality, low-cost optics and need to be constructed from simple materials using standard fabrication processes. Both designs also need to be rigid and stable enough such that wavelength drift is not a significant problem. Beyond those basic considerations, challenges become specific to each design. In brief, the compact spectrograph design is dominated by the need for low weight and maximum efficiency and signal-to-noise ratio on weak signals, while the high-dispersion spectrograph design is primarily dictated by the need to achieve high dispersion with acceptable efficiency and to minimize fabrication costs in a larger device. These considerations are addressed in detail in the following paragraphs.

Compact Research Spectrograph

While there is no shortage of potential spectroscopic research topics for amateur astronomers and astronomy students, such projects will likely be forced to involve dim objects. The brightest astronomical objects have been extensively studied, cataloged, and analyzed over the past 100 years, and little remains to be learned from them. The good news lies in the sheer numbers of dim objects. For each unit increase in the magnitude scale, there is roughly a factor of three increase in the number of objects brighter than that magnitude (e.g. there are nearly twice as many objects with brightness between magnitude 8 and 9 than there are objects that are brighter than magnitude 8). Objects brighter than magnitude 9 include over 120 000 targets, many of which have had only a simple classification performed. But in order to do research on these targets, the spectrographic systems will have to be useful at very low light levels. With this in mind, a compact research spectrograph should be designed to maximize light throughput. While maximum throughput is a common criterion for most spectrographs, it is especially critical here, and I was prepared to make sacrifices in areas such as lineshape, optical quality, and ease-of-use in order to maximize the number of photons gathered.

Comets present a special opportunity for research in spectroscopy with small telescopes. While some comets each year are periodic and thus expected, for the most part the appearance of a comet is a random event. In many cases, the light curve of a comet peaks quickly, and so time is of the essence. Amateur astronomers are well suited for studying objects of this kind. Someone, somewhere has dark skies and a free telescope to make spectroscopic observations of the evolution of the comet as it proceeds along its orbit. Professionals at observatories may or may not have the telescope time to devote to the study of these bodies on short notice. In addition, comets may be bright. In almost every year, at least one comet approaches naked eye brightness (magnitude 6), which is easily in the reach of an amateur spectroscopist with a good set-up. Since this field represents a great opportunity for research with small telescopes, I decided that the design of the compact spectrograph should be especially optimized for comet (and other diffuse object) spectroscopy.

To optimize light throughput, it is necessary to minimize the losses in the optical system. There are many places in which light can be lost in a spectrograph, including reflection losses at each surface, transmission losses through lenses, losses due to nonunity grating efficiency, vignetting, and light coupling losses. In general, the best bet is to limit the number of surfaces that the light has to reflect off or pass through, and to use coated surfaces wherever possible. Vignetting losses (i.e. surfaces blocking part of the light beam) can be minimized through optical element layout. Coupling losses are typically a big issue in small spectrographs. To minimize weight on a small telescope, a fiber optic coupling arrangement is an attractive option. However, for optimum resolution performance in a small spectrograph using fast input f numbers and commercially available grating sizes, fiber diameters in the 50 micron range (or smaller) are the most desirable, and such fibers are very lossy, with sometimes as little as 30% transmission over a two meter cable. Thus, to avoid fiber coupling and transmission losses, the first design decision was made: the spectrograph would be mounted directly on the telescope. Since additional weight on the telescope can cause balance and drive problems, minimum spectrograph weight becomes a key design criterion.

To optimize the performance of the spectrograph on diffuse objects, it is imperative to collect light from as wide a region as possible. In order to do so, the telescope must be operated at as fast an f number as possible to maximize the amount of the object that falls on the given slit size at the spectrometer inlet. For larger telescopes (8 inches and above), f/6 is a fairly fast focal ratio, though telecompressors can be purchased to bring the f number to as low as f/3.3. With this in mind, a spectrograph in the f/3 to f/4 region would seem most desirable. In addition, further increases in performance can be obtained by incorporating a longer exit focal length than the inlet focal length, resulting in an image demagnification at the spectrograph exit plane. Since the resolution in small spectrographs is typically limited by the image size of the inlet slit on the detector, a focal length demagnification can allow the use of wider slits (meaning more light gets in the spectrograph), while maintaining the same spectral resolution. The reduction in exit focal length also reduces the dispersion, however, so there is a limit to the demagnification possible before a minimum level of resolution is reached. For comet work, a resolution of 5 Å is typically sufficient to resolve and separate key features in the molecular bands. In addition, 5 Å resolution is sufficient to allow stellar classification. This level then was chosen as the maximum dispersion of the unit; lower dispersions for even dimmer objects should be attainable by switching to coarser gratings. Thus, I arrived at the following general design criteria for this instrument, which I call the Compact Research Spectrograph (CRS):

1. telescope-mounted design with minimum weight;
2. f number in f/3–f/4 range; and

3. maximum demagnification while still maintaining 5 Å resolution at highest dispersion.

High-Dispersion Spectrograph

The high-dispersion spectrograph has quite a different set of design issues. High-dispersion means operating with longer exit focal lengths and maximum groove density at the grating. Long focal lengths typically mean bigger, heavier spectrographs, so telescope mounted designs are probably not the way to go for a small telescope when high dispersion is needed. In this case, I was forced to go with an optical fiber coupling scheme between the telescope and spectrograph.

What is a desirable level of dispersion? To resolve any structure in the planetary absorption bands (e.g. CO_2, CH_4, NH_3, H_2) in Venus, Saturn, and Jupiter, from which temperature estimates can be made, a resolution in the neighborhood of 0.5 Å is required. If three-pixel resolution is attained (assuming 10 micron pixels), 0.5 Å resolution requires a dispersion of at least 17 Å/mm. With an 1800 gr/mm grating, such dispersion necessitates a focal length in the neighborhood of 300 mm. To be safe then, I figured a focal length somewhat larger than 300 mm would be best. This allowed some room for error if the three-pixel resolution is not attainable.

To accommodate tracking errors at the focal plane as well as to minimize fiber transmission losses, I wanted to choose a larger fiber. To offer a variable resolution for a fixed fiber diameter, I chose to have the fiber butt up against a variable slit at the inlet, allowing a boost in resolution at the expense of signal, by narrowing the inlet slit.

The f number of the spectrograph should not be larger than 10, since that is near the upper limit on common small telescopes, and f/10 is also a common ratio for 8-inch, 10-inch, and 12-inch Schmidt–Cassegrain telescopes. Slower optics will result in light losses that scale as f number squared. Since there already will be light loss due to focal ratio degradation in the fiber, f/10 seemed like a reasonable upper limit.

Thus the design criteria for the high-dispersion spectrograph (HDS) are:

1. exit focal length greater than 300 mm (to achieve 0.5 Å resolution);

2. fiber-coupled design with a variable inlet slit; and
3. f number no greater than f/10.

Spectrograph Design and Construction

Within the given design criteria, there is still plenty of room for different design. The designs that I converged upon were certainly not the only possible solutions, but they represent what appeared at the time to be the best compromise when factors such as ease of construction were considered. The sections below detail the explicit choices of optics, optical configuration, and overall spectrograph geometry.

Compact Research Spectrograph

To achieve the stated 5 Å maximum resolution in the CRS, I started the design process by assuming 2.5 pixel resolution (again using 10 micron pixels) and a grating density of 1800 grooves/mm. Two pixels is about the best resolution attainable under any circumstances, so 2.5 pixels seemed like a reasonable goal. Higher groove densities than 1800 grooves/mm are available, but they typically require very large grating angles, resulting in reduced projected areas and beam occlusion. Using these two criteria and estimating a ballpark deviation angle of 30°, I came up with a required exit focal length of around 24 mm. As a focusing element, a camera lens is typically a good choice, since focusing is easy, and optical quality is good even at fast f numbers. Since I wanted a demagnification, fast exit optics were essential. After looking at available inexpensive camera lenses, I chose a 25 mm f/1.3 CCD camera lens for US $75 as the exit optic for the CRS.

The f/1.3 camera lens has an opening of 19 mm, so collimated beams of less than 15 mm diameter or so will be vignetted for grating to lens distances of a few centimeters. Thus, I chose 12.5 mm (~1/2 inch) as the diameter of the collimated beam. A 12.5 mm collimated beam fits nicely on a 1-inch square grating, even at fairly high incidence angles, so I was able to use this standard grating size.

The remaining optics selection concerned the inlet focal length and f number. To achieve the design criteria, I could have chosen an inlet focal length of 38 mm (f/3) yielding a demagnification (entrance focal length/exit focal length) of 1.5. A 50 mm entrance focal length (f/4) yields a demagnification of 2.0. On the basis of pure signal at a fixed 2.5 pixel resolution in a diffuse object, the f/3 configuration would supply more signal that the f/4 configuration by a ratio of 4:3. This ratio is a result of the fact that the light gathered for a given slit size scales as $(f/)2$, but the required slit width to achieve a specified level of dispersion for a fixed exit focal length scales as $(f/)-1$. However, the f/3 configuration would require a 38 micron slit to achieve 2.5 pixel resolution, and 38 microns is a nonstandard size for fixed slits. In addition, when operating the telescope at higher f numbers – as would most often be the case since f numbers below f/4 are hard to achieve with most large-aperture commercial telescopes – the 38 micron slit covers only 4–6 arc seconds (8-inch telescope at f/10–f/6.3). This range approaches the resolution limit under poor observing conditions, and thus light will be lost at the slit unless tracking is perfect. The 50 micron slit required by the f/4 configuration is a standard size and gives a bit more room for tracking errors and imperfect seeing. With this compromise in mind, I chose the f/4 configuration which made the inlet focal length 50 mm.

The optical path for the CRS was then designed around the chosen set of optics. For a focusing element, I could have used a reflective element (mirror), but transmissive optics often offer more compact configurations for fast spectrographs, and achromats can be used to correct for on-axis spherical aberrations. Lenses are also commercially available in a wider variety of sizes and focal lengths. With this in mind, I chose a 15 mm diameter, 50 mm focal length coated achromat as the focusing element. To minimize aberrations, I wanted to keep the deviation angle (Dv) as low as possible. I'd used designs with Dv = 30° before, and that seemed about the minimum to accommodate different exit and entrance optics without occluding the collimated beam. Sketches of potential optical configurations for this spectrograph confirmed that assertion, so I stuck with Dv = 30°.

To keep the weight of the camera close to the telescope (in order to minimize the moment on the mount), I sketched out several configurations for turning the inlet or collimated beam so that the exit beam would be close to perpendicular to the inlet. Figure 7.1 shows the configuration that I finally decided upon. In order to achieve the compact design, I had to include the small coated turning mirror right after the slit. Since the mirror surface has greater than 95% reflectivity, I felt that this added loss was acceptable.

Figure 7.1. A sketch of the Compact Research Spectrograph (CRS). The labeled parts are as follows: A – 1.25-inch inlet; B – fixed slit assembly; C – turning mirror; D – 50 mm f/l, 15 mm diameter achromat; E – grating; F – micrometer for grating angle adjustment; G – f/1.3, 25 mm f/l camera lens; H – CCD camera mount.

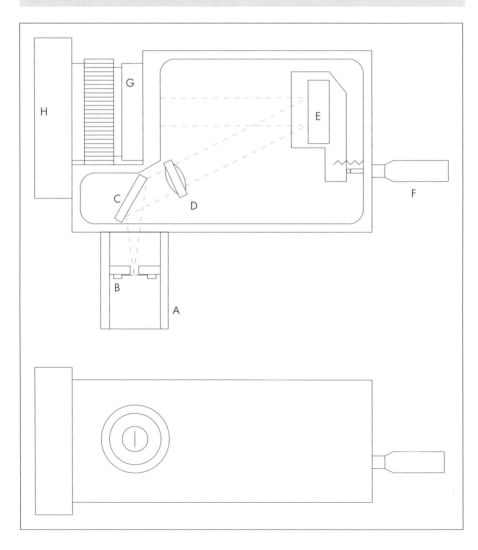

Once the overall configuration was selected, the details of the CRS's construction were finalized. In general, good design involves thinking about fabrication at an early stage in the design process, and that was the case here. Indeed some design decisions, such as the choice of fixed instead of variable slits and the perpendicular inlet and outlet beams, were significantly affected by ease-of-manufacturing considerations. For the final construction, I chose to make the spectrograph body out of a single piece of aluminum, which has essentially the same stiffness to weight ratio as steel, and yet is cheaper and easier to machine. By carving out the spectrograph body from a block, considerable time is saved since fewer pieces have to be made, and fewer set-ups on the machine are required. I selected wall thicknesses of 0.2 inches, since this would retain enough stiffness to minimize flexure, but would not add excessive weight. The mirror, achromat, and grating mounts were all to be made from standard sheet (3/16 inches thick), again to minimize the number of fabrication steps required.

The camera lens was mounted external to the spectrograph to minimize weight and allow focusing without opening up the spectrograph cover. An external plate secured the lens in place and also served as the coupler for the camera. To rotate the grating, I used a simple micrometer that pushed on the plate that held the grating. The grating plate was connected to a short shaft that protruded through the spectrograph bottom. The grating thus could rotate freely when pushed on by the micrometer. A small spring maintained the contact between micrometer and grating plate, allowing smooth rotation in both directions.

To allow for interchange of gratings and a large range of grating angles in both positive and negative orders, I drilled and tapped a series of holes for the post assembly that pushes down on the top of the grating to keep it in place. With this set-up, incidence angles from $+90°$ to $-60°$ (i.e. the maximum possible range that will produce a spectrum with $Dv = 30°$) are allowed.

All components were black anodized to minimize stray light. Painting the surfaces would have been superior, but paint can interfere with mating and moving parts in some cases, I stuck with the anodizing. The spectrograph cover was cut from a 1/16-inch sheet, and a layer of thin, adhesive-backed velvet was used to make the light seal.

Figure 7.2 shows a photograph of the spectrograph. This unit cost approximately $400 ($100 grating, $70 camera lens, $40 achromat, $20 turning mirror, $50 slit, $50 micrometer, $70 aluminum), not including the cost of machining. I did the machining myself, and the time involved was approximately 20 hours.

High-Dispersion Spectrograph

The task of design and construction of the HDS was undertaken in conjunction with my undergraduate student Alyssa Sandrowitz. Since we would be doing the construction ourselves, simplicity of design and construction were essential. High tolerances and complex geometries were to be avoided.

A few basic issues limit the design of the HDS. The largest grating that is commercially available from stock is 50 mm square. Larger sizes can be purchased, but at roughly 10 times the cost (>$1000), so the 50 mm square grating was chosen. This limits the beam size to below 50 mm. For an f/10 spectrograph, the maximum inlet focal length would then be 500 mm. Large focal length f/10 spectrographs typically are of either Czerny–Turner or Fastie–Ebert designs, since these are the simplest configurations. The Czerny–Turner involves two mirrors and the Fastie–Ebert one large one. Both spectrograph layouts at f/10 are long and narrow. Looking at available optics, the Czerny–Turner design seemed easiest to implement, since no commercially available mirror seemed suitable for a Fastie–Ebert

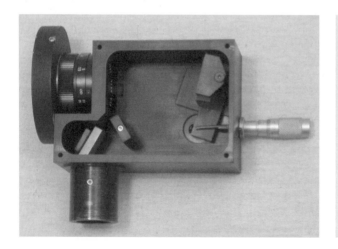

Figure 7.2. A photograph of the CRS with the cover removed.

design. For a potential Czerny–Turner design, we found 444 mm focal length parabolic mirrors from Edmund Scientific for less than $70 apiece, and these seemed very well suited to this task. The parabolic mirrors would provide some off-axis astigmatism, but in a spectrograph this is not necessarily a problem since it simply leads to a widening of the spectrum. Sketches of the layout suggested that a compact design with a small Dv = 14° was attainable, so this was chosen as the deviation angle.

The grating mount was similar to that used in the compact design, only larger to accommodate the 50 mm square grating. The grating rotation was induced by a 1-inch travel micrometer that pushed on an arm that attached to the grating turret. The arm could be attached to one of several holes in the turret, again allowing for the use of different gratings in different orders.

To mount the large mirrors, we stuck with commercial mounts for the 3-inch diameter mirrors, again from Edmund at $90 apiece. These allowed for secure mounting and fine adjustment and represented a cost savings over making the mounts in-house.

The inlet and outlet light ports followed simple designs. For the outlet, where the camera would be positioned, we simply specified a tightly toleranced hole in a 1-inch block where a T-mount barrel could be placed and secured by a nylon-tipped set screw. The 1-inch thickness and 0.005-inch clearance kept the camera mount from moving around too much as the focus was obtained by sliding the barrel in or out of the hole. The inlet would be the standard SMA fiber-optic mount, and we purchased a ferrule and jam nut for this purpose, and had the ferrule thread into the standard $1/4$-36 tapped hole. At the back end of the mount, the SMA ferrule butted up against a pair of slits held in place by two screws in rather loose clearance holes. The slop in the holes allowed the slits to be adjusted over a range of 0 to 500 microns. The slits were made from aluminum sheet that was sharpened by hand using a file and then fine sandpaper. The final finish on the edge would certainly not be excellent over long areas, but over the few hundred microns of the fiber diameter, there would not be a problem. The entire inlet assembly (ferrule, jam nut, threaded hole, and slit assembly) was mounted on a 1-inch diameter plug that fitted into a 1-inch hole in the front wall of the spectrograph. The plug could be easily removed and replaced for slit adjustments as required.

The connector for the telescope end of the fiber optic cable was similar to that at the spectrograph end, only the diameter was 1-1/4 inches (telescope standard), and there was no slit assembly. The design called for the fiber end to sit right at the exit plane of the telescope's 1-1/4-inch bore, so that the focus would be identical to that of most eyepieces. Thus, for operation, the desired astronomical object could be centered and focused in an eyepiece, then the eyepiece replaced with the fiber coupler, and no further focus adjustment should be necessary.

Due to the large size, carving out a cavity for this spectrometer from a solid block was out of the question. We would have to make all the surfaces from plate. For the base, we chose $^1/_2$-inch aluminum plate that would be stiff enough to maintain the relative positions of the optics under mild stresses associated with moving the spectrograph around or setting it on various surfaces. For the walls, we chose standard stock sizes (1/4 inch × 4 inches) and left the width as-is so that only two ends had to be machined, and the holes to attach the pieces drilled and tapped. This choice cut the machining operations in half. The front panels with the 1-inch thick pieces were also cut from standard 1-inch × 4-inch bar. The top was made from 1/8-inch sheet, and again employed an adhesive backed velvet light seal.

We chose to paint rather than anodize this spectrograph. The larger surfaces and the fact that we could disassemble the unit and didn't have to spray paint inside cavities made the process simpler than for smaller spectrographs. Preliminary testing showed the need for some baffling to eliminate stray light. Baffles were easily constructed from sheet metal and stiff paper, and could be taped in place using photographic tape. The use of three baffles cut stray light down by more than an order of magnitude without affecting throughput.

A sketch and photograph of the finished spectrograph are shown in Figs 7.3 and 7.4, with the key elements identified in Fig. 7.3. The total cost (excluding the simple machining processes) was: $100 aluminum + $50 micrometer + $150 grating + $140 mirrors + $180 mounts + $100 fiber = $720. Construction involved roughly 20 hours of simple operations in the machine shop.

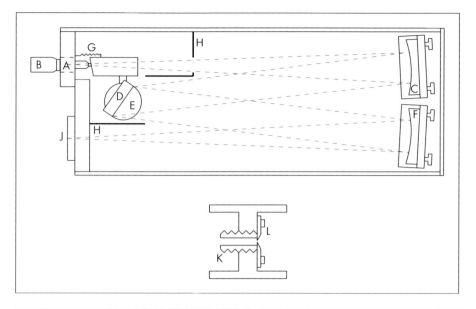

Figure 7.3. A sketch of the high-dispersion spectrograph (HDS). The item below the spectrograph sketch is a close-up of the inlet assembly. The labeled parts are as follows: A – inlet; B – micrometer for grating angle adjustment; C – collimating mirror; D – grating; E – grating turret; F – focusing mirror; G – return spring; H – baffles; J – CCD mount; K – SMA connector; L – variable slit assembly.

Performance

Before any field testing was attempted, each of the spectrographs was bench-tested using standard calibration sources for linewidth (FWHM) and lineshape. The spectrographs were also tested for day-to-day wave-

Figure 7.4. A photograph of the HDS (cover removed) with the telescope adaptor and fiber optic cable.

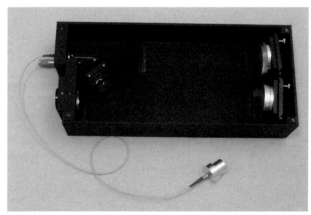

length drift and performance in the near-UV (down to 3500 Å) and near-IR (up to 9000 Å). Since transmissive optics were used in the CRS, the change in focus settings as a function of wavelength was also evaluated, as was the lineshape as a function of wavelength for a fixed focus setting.

Compact Research Spectrograph

The CRS was tested with an 1800 gr/mm grating to provide performance specifications near the highest reasonable level of dispersion. Figure 7.5 shows the green portion of the spectrum of a neon calibration lamp, as imaged by the CRS with a 50 micron slit. The image compression yields line images that have a full width at half-maximum of 25 microns, or 2.5 pixels. The lines are clean and fairly symmetric, showing little evidence of coma, which would appear as an asymmetric bump on one side of the spectral lines. The green doublet of neon, with a separation of 10.3 Å is clearly resolved, and the peaks entirely separated by a valley that drops to the level of the background. The measured

Figure 7.5. A neon calibration spectrum taken in the green with the CRS. The spectrum shows that the desired 2.5 pixel resolution with a 50 micron slit and 1800 gr/mm grating was achieved.

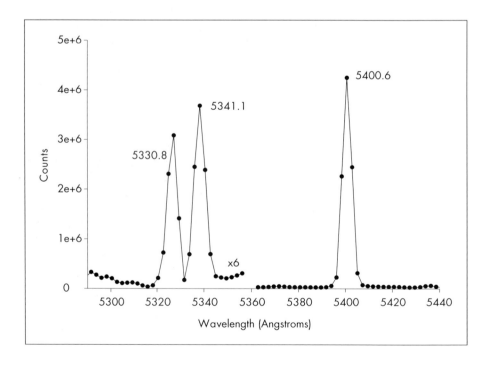

dispersion in the green is 220 Å/mm, so that 2.5 pixels corresponds to a resolution of 5.6 Å – very close to my desired 5 Å target.

In scanning from the blue to the red, it was found that slightly different focus settings yielded the sharpest images. Since the achromats and camera lenses are typically optimized for wavelengths from 4000 to 7000 Å, only small changes in focus were expected in that range. However, in the optical configuration of a spectrograph with transmissive inlet and exit optics, chromatic shifts in focal length are additive, so that errors are doubled. Even in the 4000–7000 Å range, optimization of the focus significantly improved resolution. In some regions, a degradation of linewidth was observed from the center of the chip to the edge, though careful focusing could typically keep the FWHM of lines below three pixels across the entire chip. CCDs used for astronomy are typically sensitive down to at least 3800 Å and out to 9000 Å or more. To probe outside the visible wavelengths, I found that more significant focus changes were required, and in the near-UV and near-IR, even achieving three-pixel resolution was difficult, and larger changes in linewidth across the chip were observed.

Wavelength drift is always an issue in spectrographs and can be minimized to some extent by careful control over the temperature of the optical systems and/or by choosing materials with small coefficients of thermal expansion. Neither option was available for the CRS, so I simply had to evaluate the magnitude of the drift to see if it was acceptable. This test was best performed in the field with a spectrograph that had plenty of time to thermally equilibrate. I took calibration frames before and after a long exposure (~15 minutes). The agreement between the two calibration frames was always well within a pixel

At certain wavelength settings, re-entrant spectra were observed. Such spectra can arise from light that reflects off the window in front of the CCD, then back off the grating to the chip. This is a common problem when the grating normal is close to the exit optical axis. Due to manufacturing imperfections, the spectrum rarely lies exactly on the chip centerline, and actually approaches the chip at a small angle. This small angle actually assists in minimizing the effects of re-entrant spectra since it shifts the re-entrant spectrum vertically with respect to the actual spectrum, allowing the user to easily discriminate between the two. I observed re-entrant spectra with an 1800 gr/mm grating in a 300 Å range around a center wavelength of 4300 Å. For all cases, the spectrum was shifted enough to eliminate interference with the actual spectrum, so there was not a problem.

The CRS performed on the bench pretty much as expected. Bench tests highlighted the need for accurate focus for each wavelength range, but otherwise suggested that this was a very robust instrument for spectroscopy.

High-Dispersion Spectrograph

The HDS was tested under a variety of different conditions. Since the focusing mechanism for the HDS was coarser than that of the CRS (the camera was moved manually in or out of the mounting barrel), we were concerned with how precise the focus had to be in order to obtain good resolution. Fortunately, with the slow optics and long focal length, the focus turned out to be not too critical at all. Changes of half a millimeter about the optimum focus produced negligible changes in the line profiles.

Astigmatism was, as expected, clearly visible in the spectrum. At optimum focus for minimum linewidth, the image height was significantly larger than the 400 micron fiber diameter. However, since we were not concerned with any variations in the light intensity or spectral content across the fiber diameter, the astigmatism did not result in a loss in performance.

Of primary importance with the HDS was the level of resolution attainable. With a 400 micron fiber and the slits set at roughly 30 to 50 microns and a 2400 gr/mm grating, spectra of a xenon discharge were taken in the vicinity of the doublet near 5820 Å. This xenon doublet has a separation of 0.91 Å, and thus is an excellent test for true subangstrom resolution. The resulting spectrum is shown in Fig. 7.6. The two peaks are clearly resolved and separated by a minimum that reaches the background level. The FWHM of the two lines is just slightly less than five pixels. At the observed dispersion of 0.0674 Å/pixel (6.74 Å/mm), the spectral resolution is thus 0.34 Å, which is well below our desired 0.5 Å resolution required for planetary studies.

A simple test in absorption with an 1800 gr/mm grating providing a dispersion of 7.5 Å/mm was used to test the spectrograph for very small slit sizes. I used a single-mode fiber with a diameter of eight microns to

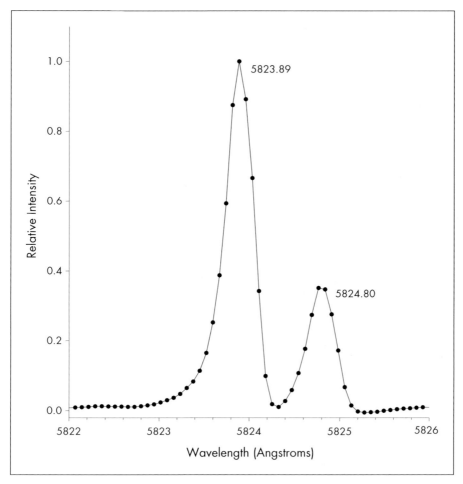

Figure 7.6. A xenon calibration spectrum taken with the HDS. This doublet is separated by only 0.91 Å, and it is easily resolved and split with the HDS.

provide the input, and simply pointed the other end at the Sun during the day. The resulting spectrum in the neighborhood of the sodium D doublet is shown in Fig. 7.7. Note the wide separation of the D_1 and D_2 lines, and the multitude of water vapor lines in the vicinity. The resolution here was approximately 0.2 Å, or roughly 2.7 pixels, which is likely very close to the limiting resolution in this instrument.

The line profiles showed some evidence of coma, which is seen in the asymmetric line profile with the blue side of the line sloping more gradually at the base than the red side. Based on our observations of line profiles of other spectrometers (notably those of Fastie–Ebert designs), these line profiles seemed comparable.

Similar wavelength drift measurements were done with the HDS as with the CRS. Again the result was the

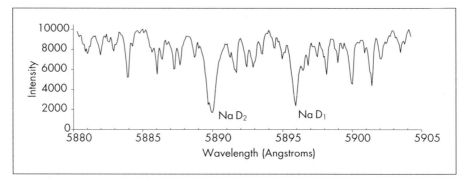

same: wavelength drift was minimal, as long as calibration frames were taken very close in time to the actual spectrum. Since all the optics in the HDS are reflective, no changes in focus were required as wavelength settings were changed.

Operation in the Field

Compact Research Spectrograph

Due to its telescope-mounted design, I expected the CRS to be much more difficult to use in the field than the HDS. In actuality, its operation was only slightly more complicated. To begin the set-up, the focus of the telescope had to be set to the location of the entrance slit of the spectrograph. When fully inserted into the 1-1/4-inch eyepiece mount of the telescope, the slit actually sits roughly $\frac{1}{2}$ inch below the exit plane of the eyepiece mount. Since most eyepieces have focal planes that are either at the exit plane or outside it, I constructed a simple adaptor that let a 1-inch eyepiece sit recessed in a 1-1/4-inch eyepiece mount. Thus, the focal plane of this eyepiece would then be close to the location of the slit. Once the telescope was focused with the special eyepiece, I could insert the spectrograph, and the focus would be at least close to the optimal value.

To fine-tune the focus, the grating was set to zero order, so that undispersed light would be visible at the exit plane. Instead of using a camera at the spectrometer exit, I looked into the spectrometer from the exit side using an eyepiece on a simple

Figure 7.7. A solar spectrum in the vicinity of the sodium doublet taken with the HDS. The HDS not only resolves the doublet, but also resolves a series of water vapor absorption lines between Na D_1 and D_2.

adaptor. I pointed the telescope (using the guide scope) at a bright star in the vicinity of the object that I want to analyze, then moved the telescope on both axes while looking at the image at the spectrograph exit until this object appears in the center of the slit. At that point, I could both optimize focus, as well as alignment of the object. Once the focus and alignment were established, then the position of the object in the guide scope was noted. This position was maintained during the exposure of the spectrum. To insure maximum precision, the longest focal length guide scope and highest power reticle eyepiece possible are used. My set-up includes a f/5.6 90 mm Maksutov–Cassegrain with a $2\times$ Barlow for an effective focal length of 1 meter, and my reticle eyepiece has a short 9 mm focal length for a magnification of $112\times$.

To take the spectrum, the eyepiece is replaced with the CCD camera, and the appropriate wavelength range dialed in on the micrometer. The telescope can be moved to the object of interest and adjusted until that object sits at the correct position in the guide scope reticle. At that point, the exposure can begin. Since the periodic error of most telescopes is enough to move most objects in and out of the slit during exposures, some manual guiding is still necessary, though a proper polar alignment goes a long way towards minimizing the frequency of adjustments.

With this general approach, I was able to take spectra of a wide variety of diffuse and point objects. The limiting factor was often the ability to see the object in the guide scope. This limitation was only a function of the guide scope and skyglow, mostly the latter. Still, spectra of objects to magnitude 9 were taken without major difficulty.

High-Dispersion Spectrograph

As mentioned above, the operation of the HDS in the field was slightly simpler than the operation of the CRS. As with the CRS, a high magnification guide scope was essential. Focusing alignment was much more easily achieved, since the fiber coupler put the fiber end right at the exit to the 1-1/4-inch eyepiece mount of the telescope. So to get the focus right, I simply pointed the telescope at a bright star and focused it using a standard eyepiece. To get the rest of the alignment correct involved a slightly modified procedure than that used with the CRS. First, since the HDS spectrograph is not an imaging spectrograph, the zero-order image will only change in brightness, not shape, as alignment is optimized. Thus, determining the optimum alignment is more challenging. Second, the grating turret in the HDS is not designed for large rotations, but rather for small, high-precision rotations. So zero order is not accessible in this instrument anyway. To work around these limitations, I made a light intensity meter from a simple photodiode that could output to a voltmeter. Once I had the focus set and the fiber coupler mounted in the telescope, I disconnected the fiber at the inlet to the spectrometer and inserted it into my rudimentary light meter. Then I monitored the light intensity as I moved the telescope along both axes. It was quite easy to discern when the fiber was optimally aligned, and then the correct position of the object in the guide scope could be established.

After this alignment, the operation was fairly straightforward. The fiber was reinserted into the spectrometer inlet, and the spectrum exposure taken while using the guide scope to maintain alignment. Using this technique, alignments could be performed in a minute or two, and high-dispersion spectra of the bright stars and planets were readily obtained thereafter.

The Spectra

Spectra with the CRS and HDS were taken over a period of several months. All these spectra were taken with an 8-inch Schmidt–Cassegrain telescope. For many of the spectra, I didn't even bother to go outside. I simply opened a window in my house or garage and collected data. For some of the work with the HDS, I didn't even bother to open the window. While such an approach would yield very poor results for imaging – due to thermal convection currents at building walls causing distortion – image quality is only a secondary concern in spectroscopy. As long as the image fits inside the slit, it can be clean or distorted, and the spectrum will essentially be the same.

Compact Research Spectrograph

Some early tests of the CRS were aimed at testing the performance of the unit in limiting cases. To begin with, I was interested in how far into the UV this device would yield useful information. Several issues work against UV spectroscopy with this conventional equipment. These issues include: decreasing CCD efficiency, large changes in focal length with wavelength, and absorption of the light by the glass and cement used in the achromats. A simple test was performed using Vega, a type A star with a series of broad H atom absorptions in the Balmer series extending well into the UV. The spectrum is shown in Fig. 7.8. The spectrum has not been corrected for variation in efficiency with wavelength, and it is easy to see that the system efficiency is falling off rapidly in the UV. Still, lines down to at least Hθ at 3790 Å are clearly resolved, and even Hι at 3760 Å is discernable. Certainly the instrument is useful down to 3800 Å, though signal strengths in this region will be significantly reduced. One of the key comet spectrum features is the CN band near 3870 Å, which, according to this spectrogram, should be resolvable without difficulty.

Figure 7.8. The Balmer series in Vega taken with the CRS to test the near-UV performance of the spectrograph. Features down to below 3800 Å are resolvable, though the system efficiency fall-off at smaller wavelengths is pronounced.

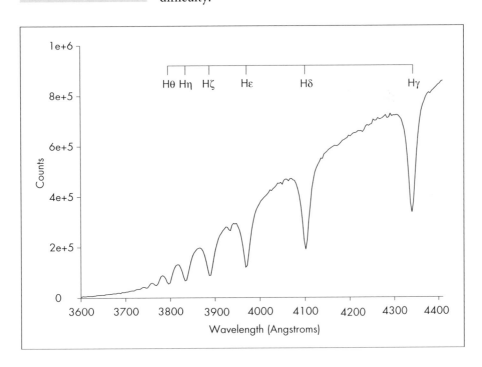

To further test the instrument, I took spectra of some stars with interesting spectra. χ Cygni is a variable star that varies from magnitude 14.2 to roughly 3 with a period of 407 days. During the period in which I was taking spectra, χ Cygni was near its maximum intensity at about magnitude 5, so I thought it a good candidate to demonstrate the spectrometer. I took an exposure of 6 minutes, focusing on the 4500–5000 Å region where the TiO bands dominate the spectrum. The resultant spectrum is shown in Fig. 7.9. Though the spectrum is fairly typical of an S class star, it seemed a waste to pass up an opportunity to capture the spectrum of this interesting variable.

Since the constellation Cygnus was in a good position for my evening observations out my den window, I moved to another unusual star: P Cygni. P Cygni is an emission line star with an expanding atmosphere around it that has emission lines that are flanked by shifted absorptions. Features of this type have become known as P Cygni features. Here I took two 8 minute exposures of this magnitude 3 star, and added them. Though I later realized that I wasn't looking in the correct place to see the P Cygni features, I was able to get a nice, high S/N ratio spectrum (see Fig. 7.10) that clearly showed several atmospheric emission lines.

Figure 7.9. A spectrum in the blue of χ Cygni when it was near magnitude 5. The sharp feature near 4800 Å is the band head of the TiO C3Δ – X3Δ, Δv = 2 absorption band.

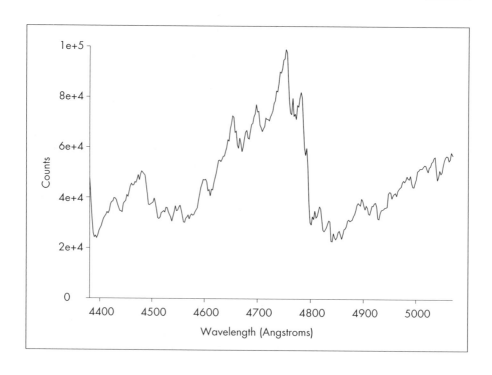

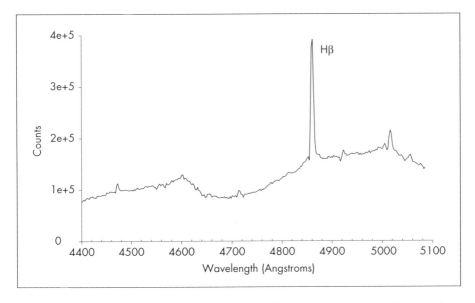

Wavelength (Angstroms)

Figure 7.10. A segment of the spectrum of P Cygnus taken with the CRS showing several emission lines including the strong Hβ line.

An absorption/emission line pair was shown more clearly in the spectrum of β Lyrae (see Fig. 7.11). Since β Lyrae is a relatively bright emission line star (at magnitude 3.5), it is a popular target for spectroscopy. The helium line near 5900 Å shows a doublet structure in emission and an absorption feature on the red side of the emission lines. The emission doublet is separated by less than 10 Å, so the CRS was easily able to resolve the individual peaks.

The dimmest object of which I took a spectrum was M57, the Ring Nebula at magnitude 9.0. Since diffuse nebulae yield line spectra (as opposed to the continuous spectra of stars), the light is concentrated in just a few discrete lines, and so higher S/N ratio spectra can be obtained on even very dim objects, provided that the object can be found and centered in the guide scope. For this observation, the nebula was barely visible in the guide scope, and it was often unclear whether it was centered or not. Nevertheless, a short 8-minute exposure (shown in Fig. 7.12) of the brightest nebula lines near 5000 Å showed the two O III lines, as well as Hβ clearly above the background.

A further opportunity to test the CRS arose in June 2000, when comet Linear S4 approached naked-eye visibility. This event provided a chance for amateur spectroscopists to put their equipment to the test. Original predictions had suggested that the comet might reach magnitude 3.5, but it actually peaked at just under magnitude 7. In order to get a spectrum of

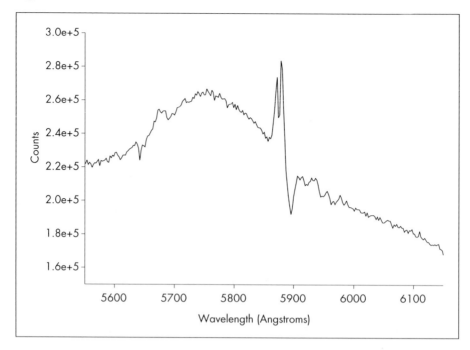

an object this dim through an 8-inch telescope, a reduced level of dispersion would be desirable, and several other amateur astronomers took this course and obtained low-resolution spectra of the comet. However, to resolve key features of the comet spectrum, I felt that at least 10 Å resolution should be achieved, and so I attempted to use the CRS with a 1200 gr/mm grating yielding 3.3 Å per pixel dispersion for a 10 Å resolution over most of the chip. Since the focus of the camera lens has a strong variation as you dip below 4000 Å, the resolution is somewhat degraded in the near-UV to roughly 20 Å around the CN band at 3880 Å.

I used a custom f/4 focal reducer to maximize the amount of the comet image that impinged upon the inlet slit. The 16-minute spectrum is shown in Fig. 7.13. A dark frame was subtracted, and skyglow was removed by subtracting the spectrum of an equal number of pixel rows, half from above the comet spectrum, and half from below. The process was incomplete on the 435 Hg line. Lines from the light pollution in the sky were also seen at 4040, 4680, 4740, 4980, and 5090 Å. The spectrum was smoothed with a triangle filter with pixel weights of 1-3-5-3-1. Spectral calibration was done using the mercury lines from skyglow.

Figure 7.11. The helium emission/ absorption feature is shown in this spectrum of ß Lyra, taken with the CRS. The emission doublet is separated by less than 10 Å, and is cleanly resolved by the CRS.

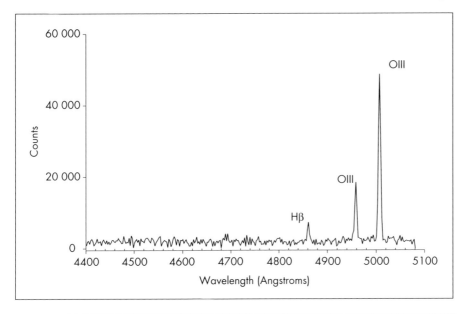

Figure 7.12. An 8-minute exposure of the ninth magnitude Ring Nebula, taken with the CRS.

The scale on the y-axis represents actual counts binned (34 10-micron pixel rows) from the spectrum. Thus, the rate of signal acquisition was only two counts per second at the strong parts of the spectrum. Despite the low signal, several key comet features are clearly resolved. These include the CN B-X band at 3880 Å and the two prominent C2 A-X Swan bands. The sharp band

Figure 7.13. A spectrum in the blue and near-UV of comet Linear S4, taken by the CRS when the comet was near magnitude 7. Several key comet spectrum features are readily identified.

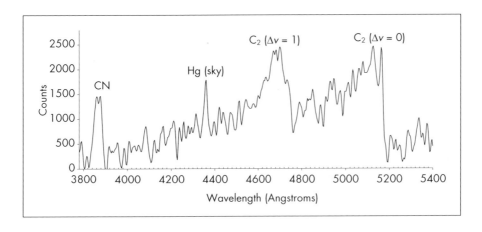

head at 5190 Å of the $\Delta v = 0$ bands is also a readily identifiable spectral signature in comets, and it is quite pronounced in the acquired spectrum.

High-Dispersion Spectrograph

The HDS was designed to resolve planetary absorption lines, so it was appropriate that it be tested by looking at the brighter planets in regions where atmospheric species produced significant absorption features. In general, there are four major species in planetary atmospheres that are readily detectable with a small telescope and spectrograph. These species are: CH_4, NH_3, and H_2 (in Jupiter, Saturn, Uranus, Neptune), and CO_2 (in Venus). Methane absorptions are very strong in the gas giants and can be easily detected at low dispersion. For this work, I looked at the latter three species and attempted to obtain highly resolved spectra of these gases using the HDS.

The general approach for looking at atmospheric absorptions is to take a spectrum of the planet first, then to do a solar comparison using, for example, a spectrum of the Moon or the Sun. Since light from the planet is primarily reflected sunlight, most planetary spectra look a lot like the solar spectrum. It is the subtle differences that yield information about the planetary atmosphere.

Carbon Dioxide

Carbon dioxide is present in the atmosphere of Venus, as was first shown in a spectrum taken in 1932 by Adams and Dunham (1932). The presence of the absorption band can be shown at low dispersion with a resolution of about 5 Å. However, to resolve individual lines in the spectrum requires much higher dispersion. Fortunately, Venus is the brightest planet in the sky, and thus there is typically plenty of light to disperse. For this work, I used the paper of Adams and Dunham as a guide, then decided to focus on the band at 7820 Å.

For my initial attempt, I used a 100 micron fiber with no slit, yielding a resolution of 2 Å. A long exposure yielded a decent spectrum, but individual rotational lines were not resolvable. For my second attempt, I switched to

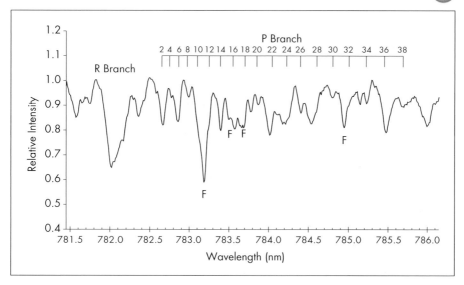

Figure 7.14. A spectrum of the planet Venus in the near-IR, taken with the HDS. Some P branch lines of the CO_2 absorption are cleanly observed, though most are blended with Earth's atmospheric water vapor absorptions.

a 400 micron fiber with a 50 micron slit. Together with an 1800 gr/mm grating, I obtained 9.7 Å/mm dispersion and a resolution of better than 0.6 Å. A 15-minute exposure yielded the spectrum shown in Fig. 7.14.

The spectrum in Fig. 7.14 shows a good signal-to-noise ratio, but does not look much like the spectra I'd seen in the literature. Several additional lines due to

Figure 7.15. A comparison of the spectrum of Venus with that of the Sun in the region of strong CO_2 absorptions.

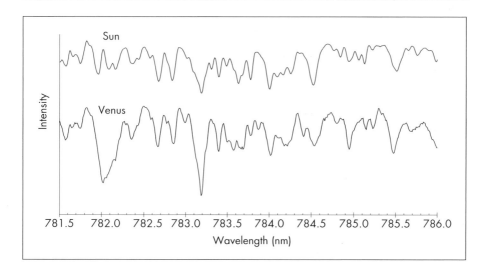

water vapor absorption in the Earth's atmosphere dominate the spectrum. This was somewhat of a disappointment, but it was also a reminder as to why observatories choose locations with dry climates.

Despite the interfering water lines, several individual rotational lines in the P branch of the CO_2 band (e.g. 8, 20, 26, and 30) are clean. A comparison of the Venus spectrum with a solar spectrum taken the next day is shown in Fig. 7.15.

Ammonia

NH_3 is also fairly abundant in the gas giants, and absorption bands exist at 5500, 6450, and 7900 Å. The 550 band is very weak, and typical CCD efficiencies are better at 6450 Å than 7900 Å, so I chose to look at the 6450 Å band. Earlier work by Spinrad and Trafton (1963) had shown that there are several ammonia absorption lines that are separated from their nearest neighbors by at least 0.9 Å. Such a band is perfect for the HDS with its subangstrom resolution.

For this exposure, I used a window in my house with a western exposure and caught Jupiter just before it set.

Figure 7.16. A spectrum of Jupiter taken in the red with the HDS showing ammonia absorption features marked by the dotted lines. Absorptions from Jupiter are broadened due to the rotation of the planet.

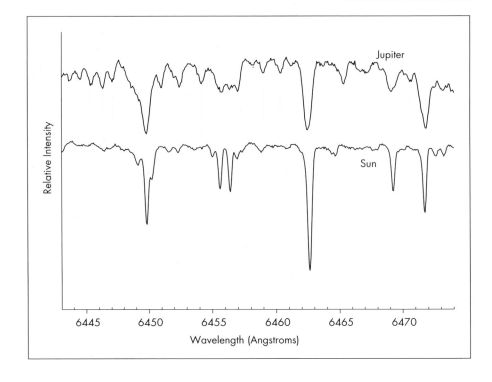

This night was very windy (with subzero wind chill), so I didn't even bother to open the window. While the object looked much fuzzier than usual through the guide scope, the poor image quality did not affect the spectroscopy. Being inside eliminated problems associated with dew formation, temperature nonuniformities, vibrations due to wind, as well as the personal discomfort of guiding for long periods in subfreezing weather.

I chose a 2400 gr/mm grating for maximum dispersion, yielding 6.6 Å/mm for a resolution of 0.35 Å using a 400 micron fiber and the 50 micron slit. A 12 minute exposure yielded plenty of signal, and the resulting spectrum is shown in Fig. 7.16, along with a solar comparison taken the next day. The broader lines in the Jupiter spectrum were puzzling at first, but then I realized that the entire planet is imaged on the fiber, and so the lines will be Doppler broadened due to the rapid rotation of the planet. There are several references to this phenomenon in the early literature on Jupiter. Despite the broadening, at least 12 individual ammonia lines are clearly resolved in this 30 Å segment of the spectrum.

Molecular Hydrogen

Molecular hydrogen (H_2) is by far the most abundant element in the atmosphere of Jupiter, comprising 90% or more of the gases in the planet (the rest is mostly helium, methane, and ammonia). However, direct detection of H_2 is difficult since it does not readily absorb visible or infrared (IR) light. The reason is that H_2, with two identical H atoms, does not have an allowed dipole absorption. (Dipole absorptions are the strongest type of absorption of light by atoms and molecules.) It does have quadrupole absorptions in the visible and IR, but these are in the range of a billion times weaker than dipole absorptions. Nevertheless, since there is so much H_2 in Jupiter, Herzberg (1938) predicted that the quadrupole absorption lines might be present, and indeed they were discovered in 1960 by Kiess, Corliss, and Kiess (1960), who identified the (3-0) lines in the near-IR around 8150 and 8270 Å. From these measurements, the amount of H_2 in Jupiter's atmosphere has been estimated. Later, Spinrad and Trafton (1963) resolved lines in the (4-0) band near 6370 Å.

At 6370 Å, I could use basically the same set-up as for ammonia (above), and I would expect similar results.

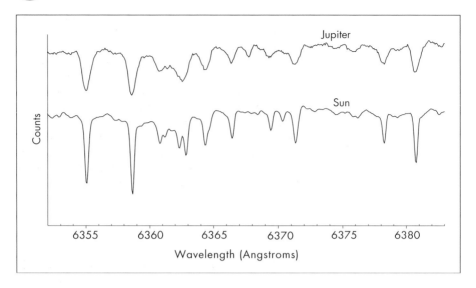

Since the S(1) line that I was looking for is weak and separated from its neighbors by only an angstrom or so, this work required high dispersion and high S/N. This spectrum, shown in Fig. 7.17, was taken outside on a warmer evening using an exposure of 15 minutes. The S(1) line was indeed weak (about 2% absorption), but at a signal-to-noise ratio of several hundred, the line was cleanly resolved and obviously absent in the solar spectrum.

Figure 7.17. The S(1) 4-0 quadrupole absorption feature of H_2 in the spectrum of the planet Jupiter is denoted by the dotted line in this HDS spectrum.

Conclusions

The two spectrographs met all the expectations set in my original guidelines. Either spectrograph could be made by an individual for less than $800 using simple shop procedures. Despite the simplicity and low cost, research-quality performance can be obtained with standard astronomical CCD cameras and small telescopes.

With a 10-inch telescope and 20-minute exposure, comet spectra out to magnitude 8 should be obtainable at 10 Å resolution using the CRS. Spectra of dimmer objects could be taken at lower resolutions. Useful data on objects out to magnitude 10 or so might be possible. The HDS can be used as a teaching tool to obtain high-dispersion spectra of planets and bright stars to clearly demonstrate and analyze planetary absorptions. While the research potential for such a unit is limited, the HDS could be used to analyze super-bright comets similar to Hale–Bopp and Hyakutake, should they appear.

In time, a larger number of commercial spectrographs will likely be available for the amateur astronomer. Should spectroscopy increase in interest and demand for spectrographs rise, then costs may come down to the level at which most individuals would rather buy than make a unit. Since we're still a long way from that right now, designs such as those I've presented in this chapter may be the most attractive alternative for the amateur astronomer or astronomy instructor wishing to delve into the area of spectroscopy.

Acknowledgements

I would like to acknowledge Alyssa Sandrowitz who did much of the design and construction work on the HDS. In addition, Shifo helped to inspire many of the designs in this work as well as my general interest in this field. Finally, my wife Patti spent countless hours assisting with the observations and proofreading the manuscript.

References

Adams, WS and Dunham, T. (1932) "Absorption Bands in the Infra-red Spectrum of Venus." *Publications of the Astronomical Society of the Pacific*, 44, p. 243.

Herzberg, G (1938) "On the Possibility of Detecting Molecular Hydrogen and Nitrogen in Planetary Atmospheres by their Rotation-Vibration Spectra." *Astrophysical Journal*, 87, p. 428.

Kiess, CC, Corliss, CH and Kiess, HK (1960) "High Dispersion Spectra of Jupiter." *Astrophysical Journal*, 132, p. 221.

Owen, T (1971) "The 5520 Å Band of Ammonia in the Spectrum of Jupiter." *Astrophysical Journal*, 164, p. 211.

Spinrad, H and Trafton, L (1963) "High Dispersion Spectra of the Outer Planets. I. Jupiter in the Visual and Red." *Icarus*, 2, p. 19.

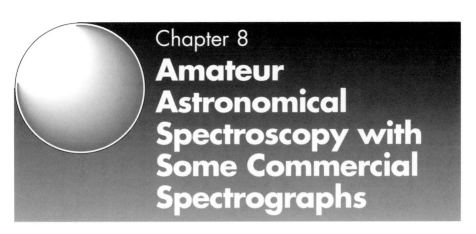

Chapter 8
Amateur Astronomical Spectroscopy with Some Commercial Spectrographs

Stephen J. Dearden

Introduction

My interest in astronomical spectroscopy goes back some 20 years, predating by about 10 years the advent of CCD imaging in the hands of amateur astronomers. At that time my simple attempts at obtaining spectra were limited to the methods that one comes across in brief sections of introductory astronomy books: the production of diffraction patterns with net curtains, the use of an objective prism in front of a tripod-mounted camera, or the use of low-quality replica diffraction gratings for capturing bright star spectra or the spectrum of a meteor. These methods are great fun to try, with some luck they can produce good results, and they are useful in wetting the appetite of the budding amateur spectroscopist. But for anyone wishing to probe further, who wants to record and, moreover, analyze the results of dispersed starlight, some type of spectroscope is necessary.[1] Prior to the early 1990s,

---

[1] I shall use the terms spectroscope, spectrometer and spectrograph interchangeably. Strictly speaking, the name spectroscope is reserved for the visual observation of spectra, spectrograph when a photographic film is the recording device and spectrometer when some form of electrical detector (e.g. CCD) is employed. All of these terms are in current usage by the professional spectroscopist.

before CCD cameras became widely available for amateurs, the detector of choice for amateur spectroscopy was the photographic film. Some excellent 35 mm camera-based spectroscopes are still around. A good example is the model 10C spectrograph manufactured by Optomechanics Research Inc. (http://www.optomechanicsresearch.com). This is an astronomical grating spectrograph that has been popular for many years with high school and college observatories for teaching as well as research purposes. Current versions of the 10C now employ CCD cameras as the detector. In addition, a few amateur-built grating and prism spectrographs using film cameras have been successfully constructed and reported in popular astronomy magazines (Sorensen 1983, 1987). Although it was tempting for me to delve deeper into these fascinating projects at that time, my educational and other commitments back in the early 1980s (I was doing graduate work at a UK university) prevented me from pursuing these interests with any vigor.

In recent years my professional job as an industrial chemist has allowed me access to spectroscopic instrumentation that is primarily designed for the research laboratory, and I have fortunately been able to test some of this equipment for astronomical applications. The instruments that I shall be describing are not usually advertised in magazines that the typical amateur astronomer reads. The reason is that this type of equipment is being targeted commercially at the industrial analytical chemistry market. The bad news is that this is a highly competitive multi-billion dollar business with much of the advertised equipment well beyond the reach financially of even the wealthiest of amateurs. The good news is that there are, fortunately, several small-sized spectrometers available from a few of these companies that do, in fact, fall within the budgets of serious amateur astronomers. These spectrometers are generally comparable in price to popular CCD cameras. This chapter describes several examples of such instruments that I have some experience with, outlines their advantages and limitations, and presents some of the results that are achievable.

The Spectroscopes

Small Hand-Held Instruments

Figure 8.1.
Small-scale fiber-optic spectrometers: (a) the Ocean Optics PC1000 spectrometer on a standard ISA interface card; (b) the Ocean Optics S2000 with a National Instruments DAQCard-700 PCMCIA card for laptop interfacing; (c) details of the S2000; the optical bench of the spectrometer is the black box on the left; (d) the SM-210 handheld spectrometer from CVI Spectral Instruments.

There are several manufacturers specializing in small-scale optical fiber spectrometers that can literally be held in the palm of your hand! Figure 8.1 shows some examples from two such companies: Ocean Optics Incorporated (Dunedin, FL, USA) and CVI Spectral Instruments (part of CVI Laser Corporation of Putnam, CT, USA). The Ocean Optics models are of two types: the PC-series is designed with the spectrometer installed directly on to a small-format ISA interface card that slots directly into a personal computer. The second type are the S-series instruments which are stand-alone versions that can be connected to a desktop computer using an interface card, or to a laptop computer via a PCMCIA card. The SM-210 spectrometer from CVI is another example of the stand-alone version.

These instruments are configured for use with optical fibers. Their primary application is for use in the analysis lab and for environmental field testing where they are used for making absorbance, reflectance and

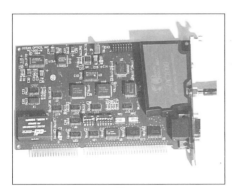

a

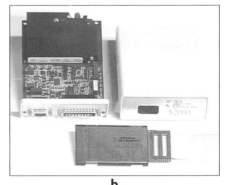

b

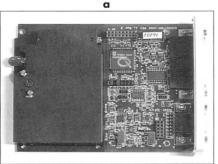

c

d

transmission measurements of a variety of materials. Their small footprint, low weight and exceptional portability make them ideally suited for these applications. For astronomy, the use of optical fibers has its advantages but also some drawbacks that I shall be discussing later in this chapter.

The PC1000 spectrometer (Fig. 8.1a) was the first-generation device sold by Ocean Optics. The unit that I have tried for astronomy has a Sony linear CCD array detector with 1024 pixels (more recent models now have 2048-pixel arrays). Ocean Optics spectrometers can be supplied with a choice of gratings, UV-visible collimating lenses and UV-visible transmitting fibers with diameters ranging from 50 m to 600 m. The gratings and lenses have to be factory-installed; they are not interchangeable and therefore the resolving power and wavelength range required must be chosen when the spectrometer is ordered. The S2000 that I am using also has a Sony linear array but with 2048 pixel elements, twice that of the PC1000. Incredibly, the whole optical bench is small enough to fit into the palm of a child's hand! The spectrometer itself is installed as a unit on a small interface card and enclosed in a metal box with dimensions of about 5 inches × 4 inches × 1.5 inches (130 × 100 × 40 mm). Prices for these instruments start around $1500–$2000 for the spectrometer, interface card, software and a couple of fibers. Cheaper versions are sometimes offered by the manufacturers for demonstration models or for to-be-discontinued equipment. For example, a recent "tag" sale that Ocean Optics advertised had some of their earlier models selling for under $1000. The interested reader should browse the Web sites of the various suppliers to see what bargains may be available (reference URLs are provided at the end of the chapter). A large range of accessories is available for fiber-optic spectrometers of this kind. These can include quartz sample holders (cuvettes), UV-visible reflection probes, immersion probes, and even a spectrometer for carrying out Raman scattering experiments. None of these accessories, however, is particularly useful for astronomical spectroscopy without some serious modifications, so any potential buyers should be aware.

Typical Spectra

Some typical spectra obtainable with these spectrometers can be seen in Fig. 8.2. This shows the recorded

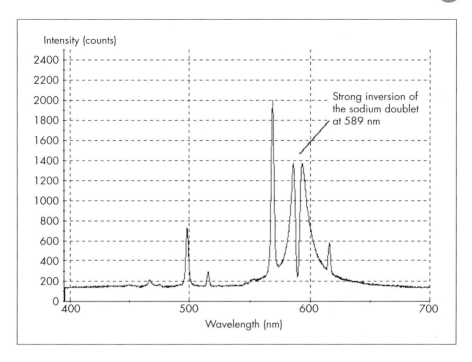

Figure 8.2. The spectrum of a local street lamp; S2000 fiber-optic spectrometer, C90 telescope, 50 micron fiber, 100 ms integration time, raw unsmoothed data with no signal averaging.

spectrum from a high-pressure sodium street lamp at a distance of about 1 km from the telescope, which was a Celestron C90. It was taken from the deck of my back garden whilst I was living in Massachusetts. A 50 μm fiber was fed between the telescope and an S2000 spectrometer. An interesting feature of the spectrum is the extremely strong inversion of the sodium doublet emission lines at 5891 Å. This rather strange line inversion is caused by self-absorption, due to photons emitted by excited Na atoms being immediately re-absorbed by neighboring atoms within the same optical path length since these lamps operate under relatively high pressures. The line is also very broad owing to pressure broadening.

As far as astronomical applications are concerned, successful results are possible, but only with objects of sufficient intrinsic brightness. Fig. 8.3 shows the reflected solar spectrum obtained by simply pointing a 100 μm optical fiber connected to the S2000 at the sky on a cloudy day. The quality is surprisingly good, with reasonable noise rejection and acceptable resolving power. The overall curved shape of the spectral profile generally reflects the spectral sensitivity of the CCD sensor, although there are other factors involved.

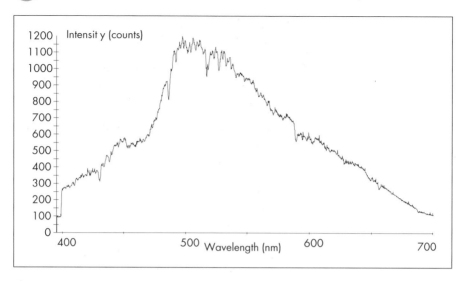

Figure 8.3. The solar spectrum, from a reflected cloud top. S2000, 100 m fiber. The Mg triplet at 5170 Å and the Na doublet at 5891 Å, although unresolved, are readily detected.

The acquisition settings in these spectrometers dictate that very short (millisecond range) exposure times are used. This is because the CCD chips used are uncooled and dark current would very rapidly increase, causing saturation if prolonged exposure times (even after several seconds duration) were attempted. A signal averaging function is available in the software that can average multiple spectra after each one has been accumulated at these short integration times. A maximum of 30 scans can be averaged in this way. The signal/noise (S/N) ratio improves by the square root of the number of scans averaged. For some lower-intensity sources this feature can help, but eventually there is a limit due to dark current accumulation.

In conclusion, I have found that these small spectrometers are attractive from the point of view of their size and weight, but their potential for regular astrospectroscopy is much more restricted. The small physical size is actually too small for much of astronomical spectroscopy, when one considers the tiny focal lengths that are being used. Optical fiber coupling is a great convenience but comes with some drawbacks. When fibers are employed, care has to be taken with f number matching (described later) which is why factory-installed collimator lenses tailored both to the spectrometer and the type of fiber are used with Ocean Optics models. The greatest drawback, in my opinion, is that the CCD chips are uncooled and thus S/N suffers considerably with the prolonged integration times necessary for astronomy. Some natural cooling of the chip may be experienced outdoors in winter which may improve matters slightly, but this is not significant and certainly does not compare to the controlled cooling achieved with multistage thermoelectric (Peltier) coolers.

A spectrum of comet Hale–Bopp has been successfully recorded with an Ocean Optics fiber-optic spectrometer and reported by Nick Glumac (http:// www.astrosurf.com/buil/us/book2.htm). Hale–Bopp, however, was a particularly bright comet. A similar spectrum obtained by this same observer but of comet Linear S4 used the more powerful Sivo Instruments Nu-View spectrograph with a cooled ST6 CCD camera. At best, small fiber-optic spectrometers such as these are limited to recording the spectra of the Sun and the brighter planets and comets. Much satisfaction and enjoyment can still be had, though, with Earth-based targets. I now use these compact instruments for recording light pollution spectra of city lights and for teaching the principles of spectroscopy to my two teenage sons!

Medium-Sized Spectrographs

The MS125™

My most useful results to date have been obtained using mid-sized spectrographs that have focal lengths of the order of 1/8 to 1/4 meter. Instruments of this size can either be directly attached to the prime focus of the telescope, or an optical fiber can be employed as a link from the telescope to the spectrograph. Figs 8.4 and 8.5 show views of a fairly typical spectrograph that can be

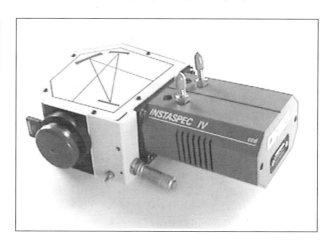

Figure 8.4. The Oriel MultiSpec MS125 spectrograph with the InstaSpec IV CCD detector head attached.

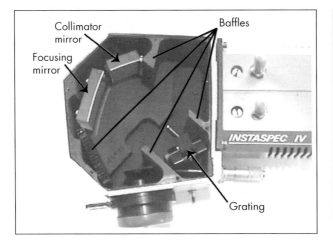

Collimator mirror

Focusing mirror

Baffles

INSTASPEC IV

Grating

Figure 8.5. Inside view of the MS125 showing internal components.

put to serious use for astronomical spectroscopy, in this case the MultiSpec MS125™ from Oriel Instruments.

This particular spectrograph is very versatile, having a range of interchangeable slits and plane reflection gratings. Some of the possible choices for gratings and slit widths, and the resulting resolution and bandpasses that are obtainable, are summarized in Tables 8.1 and 8.2.

The optical configuration employed in the MS125™ is known as a crossed Czerny–Turner mounting. With a regular Czerny–Turner spectrograph, two separate spherical mirrors are used side-by-side, and the entrance slit and detector are placed on either side of a reflection grating. The crossed design is essentially the same, but the light trains between the entrance slit and collimating mirror, and between the focusing mirror and detector, intersect each other. An advantage of this design is that it is easier to suppress stray light inside the body of the spectrograph through the use of internal baffles, and also to prevent the entrance slit from "seeing" the second (focusing) mirror, or the detector from seeing the first (collimator) mirror. The light path through this spectrograph and the positioning of the baffles are clearly shown in the schematic diagram (Fig. 8.6).

The FICS™

Another mid-sized spectrograph available from a commercial supplier is the FICS™, again from Oriel Instruments (Fig. 8.7). FICS stands for Fixed Imaging Compact Spectrograph. As the name implies, this

Table 8.1 Some Grating Specifications for the MS125 1/8 m Spectrograph (taken from Oriel Instruments Catalogue "Book of Photon Tools", p 4-80, 2000).

Lines/mm	Blaze Wavelength (nm)	Grating Type	Grating Efficiency (%)	Spectral Resolution (nm)*	Bandpass on CCD Array (nm)	Primary Wavelength Region** (nm)	Oriel Model No.
2400	400	Holographic	65	0.11	81	230–500	77420
1800	500	Holographic	80	0.14	109	300–670	77421
1200	250	Holographic	80	0.22	168	180–650	77410
1200	350	Ruled	80	0.22	170	200–1000	77411
1200	750	Ruled	80	0.22	163	450–1000	77412
600	400	Ruled	85	0.43	338	250–1300	77414
600	750	Ruled	75	0.43	338	450–2000	77415
600	1250	Ruled	85	0.43	333	750–2000	77455
400	500	Ruled	80	0.65	505	300–1200	77417
300	300	Ruled	75	0.86	668	200–750	77422

* Measured with 10μm x 2 mm slit and a 2048 pixel array.

** The primary wavelength region is where the grating efficiency is ≥20%. The system efficiency will also be affected by the reflectivity of the mirrors and the grating angle, for any wavelength.

Table 8.2 Input slit performance with the MS125

Slit Width (m)	Slit Height (mm)	Resolution (nm)*
25	3	0.4
50	3	0.6
100	3	0.9
200	3	1.4

* Measured with the 546nm Hg line, for a 1200 lpm grating and a 1024 array

spectrograph is a fixed-wavelength device since it has a permanently mounted grating. A disadvantage here is that you cannot change the spectral range of the instrument, nor can you change its resolution. On the plus side, other than an internal electronic shutter, the FICS is a very robust instrument with no moving parts to get out of alignment. My instrument has had its fair share of knocks and jolts outside on dark nights with no ill effects being observed. Another advantage of a fixed grating is that the spectrograph remains permanently calibrated. When I plan an observing session with the FICS, I simply attach the CCD detector and keep it locked in position. I then perform a wavelength calibration, as described later, and the system is then permanently calibrated until I decide to remove the detector.

The FICS instrument employs a simplified optical design that has only *two* reflecting surfaces (Fig. 8.7). These reduce reflectance losses to an absolute minimum for any spectrometer, of any design. The reflection grating

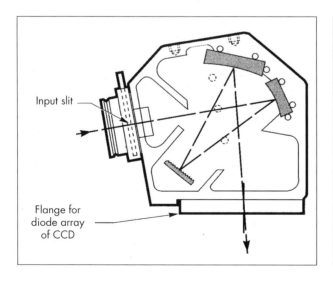

Input slit

Flange for diode array of CCD

Figure 8.6. MS125 schematic diagram. (Oriel Instruments Corporation)

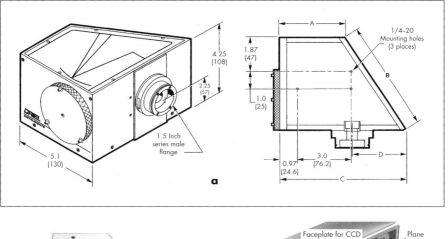

a

b

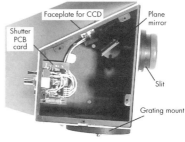

Figure 8.7. Several views of the FICS imaging spectrograph showing the simple optical design and internal detail. (8.7a: Oriel Instruments Corporation)

is a large concave ion-etched holographic type that also acts as the focusing element of the spectrograph. At f/2.1 the system is optically very fast, more than enough to capture the emergent light cone from optical fibers, a value typically around f/2.2. A big plus here is that no aperture number matching is required to prevent light overspill, and the fiber itself can act as the entrance slit.

The Detector

The MS125 and FICS spectrographs are specifically designed with linear detectors in mind for the scientific research laboratory. They are optimized for use with either photodiode arrays or charge-coupled devices (CCDs). Photodiode arrays are, as their name implies, simply detectors that consist of a linear string of tiny photodiodes. They are often found in the chemistry laboratory where they are used in UV-visible spectro-photometers, liquid chromatographs and other analytical instrumentation. The absolute sensitivity of

Figure 8.8. Close-up view of the InstaSpec IV head showing the 1024 × 64 pixel Hamamatsu sensor.

photodiode arrays, however, is much lower than that of cooled CCDs. Consequently, they are unsuitable for low light level detection required in astronomical spectroscopy. For recording the spectra of astronomical objects I have used an Oriel InstaSpec IV CCD detection system with both the MS125 and the FICS. A close-up of the CCD detector head can be seen in Fig. 8.8.

An O-ring gasket is located around the sensor window forming a light-tight seal against the mounting flange of the spectrograph. The single-stage Peltier module used with this unit is capable of lowering the temperature of the chip to 45° C below ambient conditions without assisted cooling. The two connectors at the top of the detector are for the circulation of a liquid coolant that provides an additional temperature reduction of about 10° C if required. In my set-up the sensor itself consists of a 1024 × 64 pixel array manufactured by Hamamatsu, although many other grades and formats are possible. Several characteristics of the Hamamatsu chip are shown in Table 7.3.

Table 8.3. Technical performance data of the Hamamatsu CCD

Pixel Area	24 mm^2
Total Readout noise	16.6 electrons r.m.s.*
System Gain	10 photoelectrons/count
ADC Rsolution	16 bit
Base Mean Level (Offset)	275 counts

*Total Readout Noise is all electronic noise including shot noise due to dark current. The measurement here is for one pixel (without binning) with the CCD at 25°C and an exposure time of 25 ms under dark conditions.

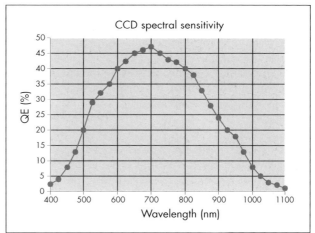

Figure 8.9. Spectral sensitivity of the Hamamatsu model 55769 CCD.

Some Comments on Blue Sensitivity

The CCD exhibits the usual fall in quantum efficiency (QE) in the blue that is typical for a regular front-illuminated CCD, and at 4000 Å the QE is down to about 2–3% (Fig. 8.9). Nevertheless, this relatively low blue sensitivity is still roughly equivalent in real terms to that obtained by fast photographic film at these wavelengths (without the attendant film problems of reciprocity failure) and can readily detect Balmer H and H lines in the blue.

In order to gain increased blue sensitivity, there are a number of options available to the spectroscopist. The easiest (but by far the most expensive!) solution is to use a backside-illuminated CCD. These are specialized chips often used by professional astronomers where the base substrate of the device has been deliberately thinned during fabrication and the photons are incident directly on the photosensitive silicon, instead of having to travel through the gate structure on the front of the chip where absorption losses occur. However, not all manufacturers offer these specialized sensors, and they are very expensive because of the additional processing stages necessary and because manufacturing yields tend to be low. Another option is to use one of the new generation of E-series CCDs from Kodak. These chips are now well known and are becoming widely available. They are manufactured with gate electrodes that are largely transparent to visible light and so significantly improve QE in the blue. In addition, they are much less expensive than the backside-thinned versions.

By far the cheapest option, though, is to consider a lumogen coating on the sensor. Lumogen is a UV phosphor material that absorbs short-wavelength photons and converts them to longer wavelength photons detectable by the CCD. (Photochemically, this mechanism is similar to the use of the so-called optical brighteners that have been used in the detergents industry for many years in order to make "whiter-than-white" soap powders!) Interestingly, lumogen is also the principal constituent used in the yellow "highlighting" pens that everyone uses these days. I personally have not yet experimented with manually applying a coating of highlighter pen solution to the CCD window for fear of damaging the sensor and affecting its performance, but this option is available to the "fearless" amateur wishing to grab more blue photons. From the technical literature, careful control of deposition conditions and coating thickness are needed to get the best effects with lumogen coatings. It should also be emphasized that only the *yellow* highlighting pens contain the correct material, which is a specific organic compound. Nevertheless, this method is, in principle, quite attractive and is by far the cheapest solution for improving blue sensitivity of the CCD. (For anyone wishing to try out this method, the interested reader is referred to Janesick and Elliot (1992) and Damento et al. (1995), but tries it entirely at his own risk!)

Data Acquisition and Processing

The CCD detector head and associated software make up the InstaSpec IV system. There are two main pixel-binning schemes that can be used with this detector: these are known as the Linear Image Sensor (LIS) mode and the MultiTrack (MT) mode. Both modes are controlled by the data acquisition software and are briefly described below.

In principle, only a single line of pixels parallel to the dispersion axis of the spectrometer's grating should be necessary to detect and record the spectrum of a particular object. In practice, many "linear" CCD sensors are not truly linear, being designed with several tens of pixels (sometimes as many as 256) in the vertical column direction as well. These are intended for use in spectral imaging applications. In MT mode, which is the imaging mode, subsets or slices of the 64-pixel columns of the Hamamatsu chip can be predefined and sent to the shift register separately, so that the CCD is effectively divided into a series of discrete tracks, again controlled by software. This capability is usually exploited in the research laboratory by employing multiple optical fibers to examine different physical areas of a light source, and then feeding the multiple signals into a single rectangular slit-shaped fiber at the entrance to the spectrograph. In this way, the slit actually samples different regions of an extended object. This is very useful for flame emission experiments or for plasma studies to spectrally examine different parts of the source. I have not yet experimented with this multitracking capability for astronomical imaging, but it may be useful for extended objects such as diffuse emission nebulae, comparing Saturn's rings and the planet's disk, or for examining different regions of planetary nebulae.

In LIS mode, each and every column of 64 elements of the 1024×64 array is vertically binned with a single transfer into the shift register of the chip prior to read-out via the amplifier. The columns are binned simultaneously and, as a result, only 1024 readouts (and hence only 1024 A/D conversions) are required, making this a very fast binning scheme that takes only 25 ms (40 Hz) per frame. Under these rapid acquisition conditions, a shutter is not required for taking spectra since there is no spatial resolution requirement in the column direction. For recording the spectra of astronomical objects I am using time-exposed conventional (nonimaging) spectroscopy with vertical binning of all 64 pixels in LIS mode to achieve very fast readout and maximum sensitivity.

Computer interfacing and data acquisition for this system is provided by a full-size 16-bit ISA card and the necessary software. The InstaSpec IV software I use is DOS based (yes, the old "dinosaur" is still around!). Nevertheless, the program provides all the required functions for data control and acquisition and also possesses a versatile control language, designed as a subset of the BASIC programming language, that allows for easy manipulation of spectral data files and even automation of data collection. For most of my recorded spectra, though, I usually convert the spectral data files from the control software's proprietary format into more portable ASCII (*.dat) format, and then onto the more convenient Windows platform. I am using GRAMS/32, which is a professional spectral data manipulation package often used for handling various spectroscopic and chromatographic data files in chemistry laboratories. GRAMS/32 is developed by Galactic Industries Corporation of Salem, NH, USA. A free demonstration disk is available or downloadable from Galactic's website (http://www.galactic.com). Two video camera images of the computer screen showing a typical acquisition are shown in Fig. 8.10.

Connecting to the Telescope

As already indicated, there are two possible ways of using mid-weight spectrographs of this kind for taking spectra: either by directly coupling the spectrograph at the prime focus of the telescope, or by using an optical fiber as a light guide between the two instruments. Both methods possess advantages as well as some drawbacks, and are described in some detail below.

The Direct Coupling Mode

This method allows the maximum possible amount of light to enter the slit of the spectrometer and therefore can be a distinct advantage compared to an optical fiber connection. Mid-sized spectrographs such as the ones described here, however, tend to weigh several pounds; when a few additional pounds due to the CCD camera are included, this can add up to a significant mass (up to 7–8 lb; 3–4 kg) at the eyepiece end of the telescope. Medium-sized telescopes from about 8 inches upward on solid mountings are therefore a necessary minimum requirement. Catadioptrics (SCTs or Maksutovs) probably offer the best compromise of size, weight and performance for directly attaching spectrographs like these. I am currently using a 12-inch (305 mm) Meade LX200 mounted on a giant field tripod in altazimuth mode with excellent results.

The principal disadvantage of the direct coupling method is the problems associated with flexure due to

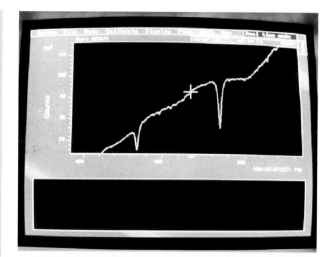

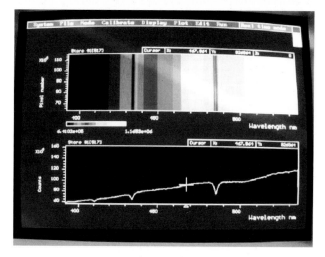

Figure 8.10. Two typical monitor screen images observed during data acquisition. The upper image, showing part of the spectrum of Sirius, is employed when operating in LIS mode. The lower image shows a false-color multitrack representation of the data above the Sirius spectrum. This is normally reserved for the spectral imaging mode (as explained in the text).

changing gravitational loadings as the telescope is pointed to different areas of the sky. The LX200's drive motors have to work hard when slewing in some directions owing to the extra load, and in particularly cold weather the telescope can respond sluggishly. But to date no problems of signal variability (which would indicate that the target image is drifting on and off the slits) have been encountered during actual guiding and exposure.

What may be surprising at first is that these results show it is not absolutely essential to use a permanently mounted equatorial telescope in a well-equipped observatory for spectroscopic work. At first glance one might feel that this ought to be a definite must. However, having moved house on several occasions in recent years due to my job, the astronomical equipment that I use has to be both mobile and portable. The LX200's computer-controlled dual-axis guiding is usually employed, and this does, of course, make life easier. But unlike long-exposure deep-sky imaging, where poor guiding or the effects of field rotation would immediately be apparent with an altazimuth mounting, the recording of astronomical spectra with my system simply requires a focused image of the object to be *maintained* on the entrance slit of the spectrometer for as long as possible. Any guiding errors, whether manually or computer generated, result only in a loss of the spectral signal that is observed in real time on the computer screen. Recovering the image back onto the slits retrieves the target spectrum again which is then tracked until adequate S/N has been accumulated. Clearly, if image shift had occurred during normal astrophotography the final result would be trailed. A "dark current" spectrum is also systematically acquired at the same exposure time but with the target object off the slit, and the two spectral data sets are subtracted, usually in GRAMS/32. An indication of my set-up, circa 2000 in Massachusetts, that shows the telescope and a directly coupled spectrometer, is shown in Figs 8.11 and 8.12.

Figure 8.11. The author, seen here with his 12-inch Meade LX200, on the backyard deck of the house in Belchertown, MA.

Optical Fiber Coupling

An optical fiber link is an attractive alternative approach for connecting the telescope and spectrometer. The clear advantage with the use of a fiber feed is that much less stress is placed on the telescope and its drive system, minimizing differential flexure and making for easier manual guiding. The obvious disadvantage is the loss in light intensity that results in transmitting light through the fiber. Most amateurs familiar with the use of optical fibers are aware of these transmission losses. But the problem does not end there. In addition, particular attention has to be paid to correct aperture matching (f number matching) at the input and output ends of the fiber, otherwise further losses in transfer efficiency will occur.

All optical fibers function on the principal of total internal reflection. If a ray of light passing through a medium of refractive index n_1 strikes the interface with a second medium of lower refractive index n_2 at an angle θ that is greater than some angle θ_c the ray will be totally reflected back into the first medium. $\theta_c = \sin^{-1}(n_2/n_1)$ and is called the critical angle. Optical fibers are designed to exploit total internal reflection by having an inner region of low refractive index (the core) surrounded by an outer sheath (the cladding) of higher refractive index (Fig. 8.13).

Light entering the fiber within a certain input cone (the acceptance cone) will be totally internally reflected at the core–cladding interface. The acceptance cone angle depends on the refractive indices of the core and cladding:

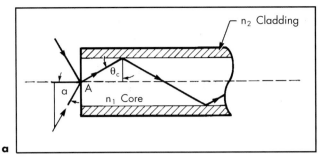

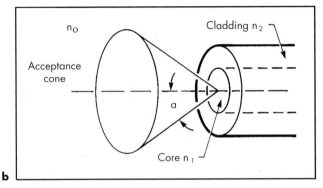

Figure 8.13. a The guiding of light within the acceptance cone through the fiber. **b** Only light entering the fiber within the acceptance cone will be totally internally reflected at the interface.

$$\sin a = \frac{1}{n_0} \sqrt{n_1^2 - n_2^2}$$

where the acceptance angle is $2a$ and n_0 is the vacuum refractive index, a number close to 1.

The calculation of transmission losses through optical fibers is a complex issue, since the values depend on the actual operating conditions and optical configuration. For example, a ray of light almost parallel to the fiber axis travels a much shorter distance than a ray that enters at the limiting acceptance angle. The reflectance at the core-cladding interface is very close to 100%, but not exactly. Thus a ray entering at a high angle can have many hundreds of reflections with even short fiber lengths and will experience more loss than an axial ray with few reflections. The overall transmission losses, therefore, are strongly dependent on the population distribution of rays forming the entrance cone. For typical situations met in amateur astronomy, fiber transmission losses of up to 40–50% can be expected.

The input and output characteristics of optical fibers are usually expressed in terms of numerical aperture, NA, which is a measure of the collecting power of an optical element. For the case of fibers here,

Table 8.4. Numerical apertures and F/numbers for typical fibers

Core	Cladding	F/number	NA	Acceptance Cone Angle 2a (deg.)
Silica	Plastic	1.9	0.27	31
Silica	Silica	2.3	0.22	25
Liquid	Plastic	1.1	0.47	56
Glass	Glass	0.9	0.56	68

$$\text{NA} = \sin a = \sqrt{n_1^2 - n_2^2}$$

NA is related to f number by

$$\text{F number} = \frac{1}{2 \sin a} = \frac{1}{2\text{NA}}$$

Values of NA and aperture number for some typical optical fibers are shown in Table 8.4.

For optimum transfer efficiency through a fiber, excluding transmission losses, the acceptance angle at the fiber input has to be wide enough to capture all of the light from the telescope. This is not usually a problem on the telescope side where instruments typically have apertures around f/10, or even when focal reducers are used that decrease the aperture to about f/5–6. The main challenge is at the spectrograph side, where the light cone entering the spectrograph must be small enough to pass through the entrance slit. If the exit light cone from the fiber is too large this will overflow the slit and waste the precious spectral signal. Overflow can even cause unwanted light scattering with some slit designs. By far the best solution at the spectrograph, and one that is well suited to linear CCDs, is to use a circular-to-rectangular fiber optic bundle (also known as a fiber shaper or image dissector). I have actually integrated a fiber shaper of this kind into my guiding technique, as explained in the next section.

Guiding with Fiber Optics

The guiding technique I use for recording spectra is currently being done manually without any electronic autoguiding equipment. I use an image dissector connected between the telescope and the spectrometer. The same basic type of device can be employed for direct coupling (Fig. 8.14) or with an optical fiber (Fig. 8.15).

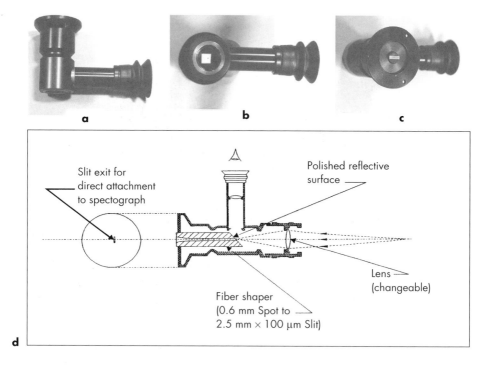

a   b   c

Slit exit for
direct attachment
to spectograph

Polished reflective
surface

Fiber shaper
(0.6 mm Spot to
2.5 mm × 100 μm Slit)

Lens
(changeable)

d

Figure 8.14. Three views of the image dissector/fiber shaper and schematic diagram. The telescope, or a standard optical fiber, replaces the camera lens shown on the left. Approximate dimensions: $4 \times 2\frac{1}{2}$ inch maximum diameter, 3-inch off-axis tube. (8.14d: Oriel Instruments Corporation)

This device is essentially a solid bundle of optical fibers of circular cross-section at one end to match a star's point image. The other end is shaped as a narrow rectangle (100 im × 2.5 mm) to match the dimensions of the entrance slit of the spectrometer. A 45 reflective surface acts as a mirror to allow visual observation of the target object. When viewed through the 90 eye-cup (Fig. 8.14), the circular end of the fiber optic appears to the eye as a small black spot in the center of the image, which accurately indicates the exact area from which the fiber gathers light and feeds it into the spectrograph. As long as a target image, such as a star, is completely covered by the guide spot, the star's image will be reshaped and accurately positioned onto the entrance slit of the spectrograph. The design allows a full-brightness image to be viewed at the same time that a spectrum is being acquired. There is no light loss that would otherwise occur if flip mirrors, off-axis prisms or beam-splitters had been used. In this way the fiber shaper gives very efficient optical coupling between telescope and spectrograph.

Originally, the device functioned as a "sighting optic" for making radiometric measurements at a distance on remote objects. It was modified slightly and a custom

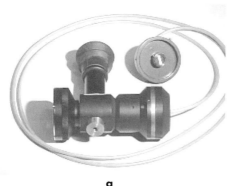
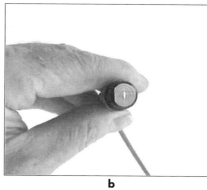

a   b

Figure 8.15. Image dissector used in combination with a fiber optic link between telescope and spectrometer. The slit-end of the fiber shaper is shown on the right.

machined coupling ring was added to connect it to the LX200. More details of image dissectors and other techniques that assist spectroscopic measurements can be found in Kitchin (1991, 1995).

Calibration Techniques

The calibration of astronomical spectra represents an important stage in the data reduction and analysis process. There are two calibration steps to consider: the conversion of the pixel number along the dispersion axis of the spectrum into wavelength units and the photometric calibration of the intensity axis. The required photometric corrections are considered briefly below and the wavelength calibration techniques that I use with my spectrographs are described afterwards.

Photometric Calibration

The y-axis of a newly acquired spectrum will be expressed in units of counts, relative intensity, or some other instrument-dependent units related to the initial measurement. Often, keeping the raw data in this form is perfectly adequate. However, if quantitative measurements of line intensities are going to be required say, to analyze the spectrum further, photometric calibration of the intensity scale becomes necessary.

The intensities of the various spectral lines in a raw spectrum certainly do not correspond to the true line intensities (in emission or absorption) of the object under observation. In practice, a large number of variables affect the real intensity of a spectral line.

These include the effects of the atmosphere, reflection and transmission factors of the telescope, the wavelength-dependent grating efficiency of the spectrometer, and the quantum efficiency and sensitivity profile of the CCD. All of these factors combine to give what is termed the instrument response function. This is a response curve unique to the telescope–spectrograph–detector system used to record the spectrum, and to the observing conditions at the time. The curve can be determined by subtracting the standard (true) spectrum of a reference star from the raw spectrum of the same object. In a sense the instrument response curve is the spectroscopic equivalent of the flat field used in regular CCD imaging. It corrects for all of the instrumental and other factors affecting the final spectral line intensities.

Complete details of photometric correction techniques for spectral data are beyond the scope and length of this chapter, and excellent descriptions already exist elsewhere. For those interested in pursuing these methods, the standard texts should be referred to (Wagner, 1992)). Other useful information can be found on the Web sites of amateurs working in the field of spectroscopy such as Christian Buil's spectroscopy pages (http://www.astrosurf.com/buil/). In addition, the Visual Spec software program, developed by Valerie Désnoux, and downloadable from her website (http://valerie.desnoux.free.fr/vspec/), has some excellent routines for performing radiometric calibration of spectral data files using a library of reference stars.

Wavelength Calibration

With CCD detection, the software automatically reads a series of digitized values corresponding to each of the pixel elements of the detector. Therefore, once acquired, spectral data are *already* in the form of intensity up the y-axis and pixel value along the grating dispersion (x) axis. However, it is often preferable to convert the horizontal axis into the more meaningful units of wavelength (λ) and this becomes absolutely essential for the spectral line (element) identification of an unknown spectrum. For astronomical spectra in the UV and visible regions of the electromagnetic spectrum, wavelength units of nanometers (nm) or angstrom units (Å) are most often used.

The calibration of my CCD array in wavelength is performed by determining a relationship between pixel number and true wavelength and is expressed as a third-order polynomial of general form:

$$\lambda_p = I + C_1 p + C_2 p^2 + C_3 p^3.$$

where λ is the wavelength of pixel number p, I is the wavelength of pixel number 0, and C_1, C_2 and C_3 are the first-order (units nm/pixel), second-order (units nm/pixel^2) and third-order (units nm/pixel^3) polynomial coefficients respectively. It is required to calculate values for the intercept I, and the three coefficients C_n.

With the spectrometers I have described, a wavelength calibration is performed using small pencil-type spectral calibration lamps of mercury vapor, argon, neon, krypton or a mixture of these gases. These are low-pressure atomic emission line sources with very narrow linewidths across the UV, visible and near IR whose wavelengths are known precisely. An example spectrum of a typical lamp of this kind, the mercury–argon, Hg(Ar), lamp is shown in Fig. 8.16. The Hg(Ar) lamp is a very useful light source for spectral calibration. It is relatively inexpensive compared to other calibration sources such as deuterium and helium lamps used by professional astronomers. It is insensitive to temperature changes, the average intensity is constant and reproducible and the lamp has a long running life. Hg(Ar) lamps typically require a couple of minutes warm-up period for the mercury vapor to dominate the discharge, then about 30 minutes to

Figure 8.16. Typical spectrum from a Hg(Ar) pencil-type calibration lamp, showing the principal emission lines in the visible.

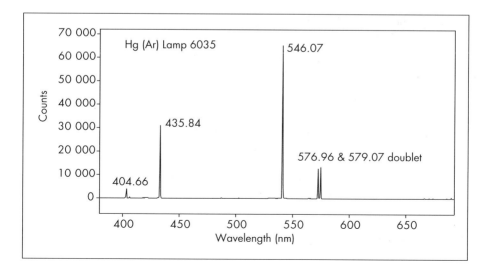

Hg (Ar) Lamp 6035

546.07

435.84

576.96 & 579.07 doublet

404.66

Counts

Wavelength (nm)

achieve complete stabilization. As a cheaper alternative you can use one of the neon indicator lights found on electrical power supplies; I have also employed these sources effectively for wavelength calibration.

With the calibration lamp, the pixel value corresponding to the exact wavelength of as many emission lines as possible in the visible is recorded. A least-squares regression analysis then evaluates the intercept and coefficients required to relate wavelength to pixel number. The goodness of fit of the regression model should be very high, and the correlation coefficient should be a number very close to 1. If it is not, this usually indicates that a wavelength value has been mistyped or incorrectly assigned during the calibration procedure. From the above analysis, the relationship between the true wavelength and the pixel number in the measured spectra is obtained.

It is important to note that the procedure is valid only for a given diffraction grating and for a given wavelength range. Should these parameters be changed, a new calibration has to be performed. Fortunately, my software stores the calibration parameters in a configuration file that is unique to the grating and λ range used. Therefore once a calibration has been carried out for a particular grating and range, it does not need repeating. I have calibration files stored on my PC's hard drive for 2400, 1800, 1200 and 600 lpm gratings, each at a series of wavelength ranges.

A practical point to note is that a least-squares fit of the pixel data is not a perfectly straight line, but exhibits some slight curvature. A second-order polynomial (a quadratic model) may have been enough to fit the data, but a third-order (cubic) model takes into account any sigmoidal (s-shaped) variation in the data set. Diffraction gratings are traditionally thought of as linear dispersing elements, and when compared to prisms they certainly come very close. However, the dispersion of a grating is not truly linear in wavelength, having small deviations that can be both in a positive sense and a negative sense. This gives rise to the slight s-shaped form of the calibration curve. The extent of nonlinear dispersion in gratings is mainly governed by the accuracy of the ruling engine that engraves the lines on the grating during fabrication. Nonlinearity is also affected by machining tolerances of the micrometer wavelength drive. Spectrographs that use micrometers to change the wavelength range (such as the MS125) use a sine-bar drive. This is a mechanism built onto the grating mount consisting of a bar with a hardened steel ball-bearing at the end in contact with the finely polished flat end of the micrometer shaft. Small variations in the machining or alignment of the sine-bar drive also contribute to the deviations from strict linearity.

More information on wavelength calibration methods and spectral calibration lamps and standard wavelengths is available on my website (http://www. astrosurf.com/dearden).

Some Results

The remaining pages of this chapter (with Figs 8.17–8.22) present some of the results that have been obtained over the last year or so with this equipment. I have deliberately avoided presenting detailed data that are beyond the intended scope of this chapter, preferring to show a representative sample of spectra of some of the brighter stars, planets and nebulae. You should note that all the spectra were acquired with relatively short exposure times of 2–4 minutes, and in one case (M42) as little as 1 minute. These are surprisingly short exposures when one considers the low light intensities involved, especially after passing through the spectrograph and dispersing across 1024

Figure 8.17. Spectrum of Sirius (spectral class A0 V) recorded on Christmas Day 2000. A 4 min exposure with the FICS spectrog raph, 50 micron slits, detector temp –20 °C. Image scale 1.4 Å/pixel.

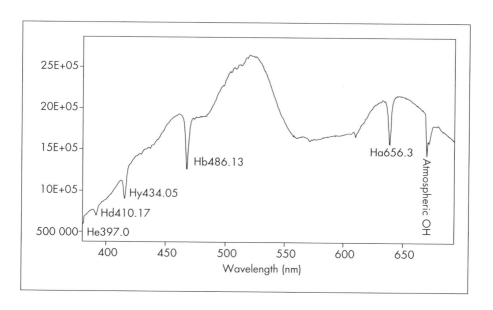

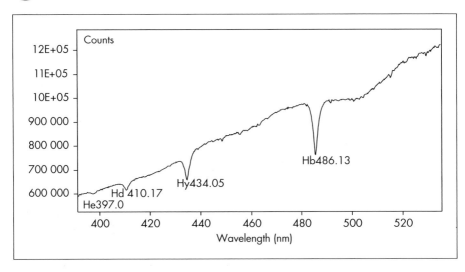

Figure 8.18. Spectrum of Sirius (spectral class A0 V). A 4 min exposure with the MS125 spectrograph, 100 micron slits, detector temp –20 °C. Image scale 1.2 Å/pixel. The loss in blue sensitivity of the CCD can be seen clearly.

pixel elements. I put this down to the efforts I paid over aperture matching and transfer efficiency, to the fast optical speed of the spectrographs, and to the good red-green sensitivity of the CCD.

Figure 8.19. The spectrum of M42, 1 minute exposure time (1200 integrations of 0.1 sec exposures), 1200 lpm grating, 100 m slit, image scale 1.4 Å/pixel.

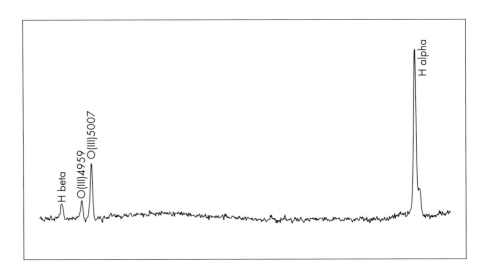

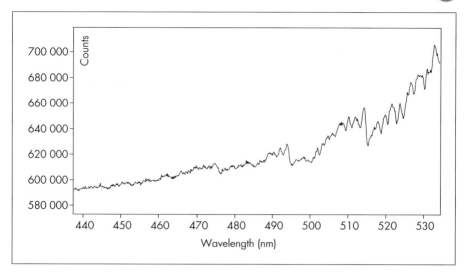

Figure 8.20. Spectrum of Betelgeuse in the blue-green region. Image scale 1.4 Å/pixel, 100 m slit, 3 min exposure time, MS125 spectrograph.

Figure 8.21. The spectrum of Jupiter, taken with the MS125 spectrograph, 2 min exposure with 100 μm slit. Most of the absorption lines are due to reflected solar radiation.

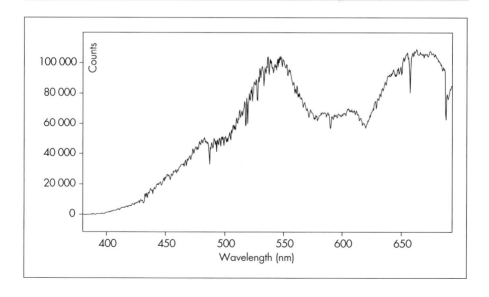

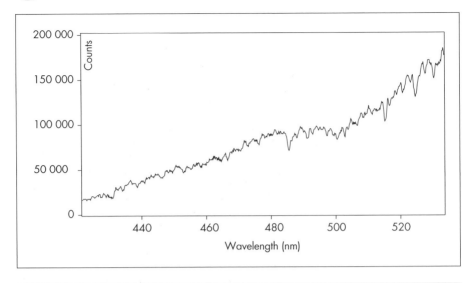

Figure 8.22. The spectrum of Capella (G8 III) in the blue-green showing a complex series of many absorption features. MS125 spectrograph, 4 min exposure, 50 m slit.

Conclusions

I have tried to give you some idea of the results that are possible with commercial equipment of this type. Their cost is comparable with what you would expect to pay these days for a good CCD camera. I myself look forward to testing the capabilities of these spectrographs in more depth, attempting longer exposure times and restricting the wavelength range with higher-dispersion gratings for greater resolution and detail. Some of the many astronomical projects that come to mind include the monitoring of P Cephei line profiles, the study of Be stars that exhibit Balmer lines in emission, or investigating the broadened emission lines from very hot Wolf–Rayet stars. Other pet projects of mine that are more Earth-bound include the spectroscopy of lightning strokes using the internal transient recording capability of the CCD detector (which I didn't go into!) and spectrophotometric studies of twilight skies after volcanic eruptions.

Although amateur astronomical spectroscopy may not turn out to be as popular as conventional CCD imaging, there is a growing number of amateurs becoming actively involved in this fascinating hobby within a hobby.

References

Buil, Christian, http://www.astrosurf.com/buil/

Damento, MA, Barcellos, AA and Schempp, WV (1995) "Stability of Lumogen Films on CCDs." SPIE Proceedings on Charge-Coupled Devices and Solid State Optical Sensors V, San José, February.

Dearden, Stephen, http://www.astrosurf.com/dearden

Désnoux, Valerie, http://valerie.desnoux.free.fr/vspec/

Galactic Industries, http://www.galactic.com

Glumac, Nick, http://www.astrosurf.com/buil/us/book2.htm

Janesick, J and Elliot, T (1992) "History and Advancements of Large Area Array Scientific CCD Imagers." In *Astronomical CCD Observing and Reduction Techniques*, ed. SB Howell, Astro. Soc. Pacific Conf. Series 23, p.1.

Kitchin, CR (1991) *Astrophysical Techniques*, 2nd edn, Adam Hilger.

Kitchin, CR (1995) *Optical Astronomical Spec*troscopy, IOP Publishing.

Sorensen, B (1983) "An Objective-Prism Spectrograph." *Sky and Telescope*, May.

Sorensen, B (1987) "A Simple Slit Spectrograph." *Sky and Telescope*, January.

Wagner, R Mark (1992) "Point Source Spectroscopy." In *Astronomical CCD Observing and Reduction Techniques*, ed. SB Howell, Astro. Soc. Pacific Conf. Series 23, p. 160.

Useful Web Sites

1. http://www.oceanoptics.com
2. http://www.cvilaser.com
3. http://www.instrumentsSA.com
4. http://www.oriel.com
5. http://www.acton-research.com
6. http://www.alcprecision.com
7. http://www.catalinasci.com

Chapter 9
Astronomical Spectroscopy with the Santa Barbara Instrument Group Self-Guiding Spectrometer

Dale E. Mais

Introduction

Spectroscopy had its beginnings in the latter half of the 19th century where it was primarily the domain of the amateur working from his private observatory. As we entered the 20th century with a greater emphasis on astrophysics, this area of research shifted toward the professional astronomer working from world-class observatories. This shift was primarily driven by the increasing costs and skills required to do state-of-the-art spectroscopy and the requirement for large telescopes due to film-based detection of a spectrum. Once again we are experiencing a shift where the amateur can make contributions to the area of spectroscopy. This is due to both the use of more sensitive CCD detectors and the recent availability of powerful and versatile spectrometers aimed at the amateur community. I will focus on the instrument produced by the Santa Barbara Instrument Group (SBIG), the Self-Guided Spectrometer (SGS). In the past, due to the limitations of film-based detection, amateurs were limited to obtaining spectra of only the brightest stars and nebulae. The SGS allows spectra to be obtained with only modest aperture instruments of stars down to 10–12th magnitude.

Equipment

My primary instrument for spectroscopy and the evaluation of the SGS is a Celestron 14, which has a Byers retrofitted drive system. The spectrometer is linked to the telescope with a focal reducer giving a final f/6 ratio. The CCD camera attached to the spectrometer is the SBIG ST-7E with 9 μm pixel size. The SGS instrument appeared on the market during the latter half of 1999 and was aimed at a subgroup of amateurs with special interest in the field of spectroscopy (www.sbig.com). The instrument is shown in Fig 9.1 with a number of features pointed out and the path of light indicated. The instrument features several very novel features. In conjunction with SBIG CCD cameras, the SGS is self-guiding in that it keeps the image of an object locked on to the entrance slit, which allows for long exposures to be taken. The light from the telescope reaches the entrance slit, which can be 18 or 72 μm wide. The light passes through the slit and reaches the grating and ultimately the CCD camera imaging chip. The remaining field of view is observed on the guiding CCD chip of the camera and allows the viewer to select a field star to guide upon once the object of interest is centered on the slit. In this chapter only results obtained using the 18 μm slit will be presented. The wider slit option allows the spectra of fainter objects to be obtained at the expense of resolution. This would be particularly of interest to those interested in measuring the red shifts of more distant and thus fainter objects since the wider slit permits an additional two magnitudes of penetration.

The SGS features a dual grating carousal, which, with the flip of a lever, allows dispersions both in the low-resolution mode (~4 Å/pixel, ~400 Å/mm) or higher-resolution mode (~1 Å/pixel, ~100 Å/mm). In the low-resolution mode, about 3000 Å coverage is obtained, whereas in the high-resolution mode, about 750 Å. The particular region of the spectrum is determined by a micrometer dial and is set by the user. The overall wavelength range of the unit is from approximately

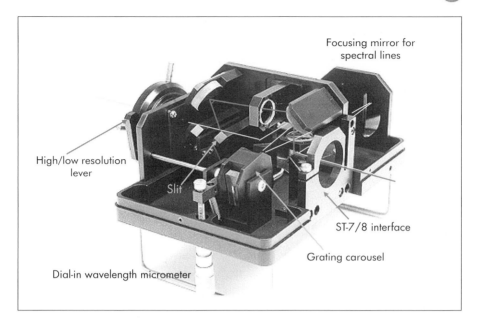

High/low resolution lever

Slit

Dial-in wavelength micrometer

Focusing mirror for spectral lines

ST-7/8 interface

Grating carousel

Figure 9.1. The Santa Barbara Instrument Group Self-Guiding Spectrometer showing various features of the instrument and the optical path. The lighter path shows the route followed by light passing through the slit to the grating and ultimately to the imaging chip of the ST-7/8 camera. The heavier path is the route followed by the light not passing through the slit and ending at the guiding chip of the same camera. (Santa Barbara Instruments Group).

10000 to 3500 Å. Spectra are obtained using CCDOPS software and are analyzed using the software package SPECTRA, both software packages from SBIG, which allows for wavelength calibration. Wavelength calibration was carried out using emission lines from hydrogen and/or mercury gas discharge tubes. These tubes, along with many others, are available from Edmund's Scientific. The light from these standard lamps is fed to the spectrometer by means of fiber optic leads into an opal window at the bottom of the spectrometer. This window has been modified slightly by the author such that the fiber optic lead remains permanently in position, as does their position at the source emission tubes. Spectral images obtained were further processed using the MaxIm software package (www.cyanogen.com/). Figure 9.2 shows the general procedure utilized to obtain and process a spectrum. The initial spectrum consists of a swath of light only a few pixels wide. Following dark subtraction, the image is cropped to a 765 × 20 pixel area, which allows it to now be imported to SPECTRA for further analysis, which includes wavelength calibration and expansion into the more familiar "line spectrum". In addition, a calibrated text file can be saved and imported into any one of a variety of graphic spreadsheets such as Excel, where the flux calibration can be carried out. Once the spectrum is expanded into a line profile, various

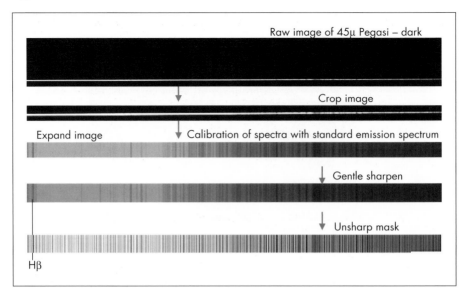

Raw image of 45µ Pegasi – dark

Crop image

Expand image   Calibration of spectra with standard emission spectrum

Gentle sharpen

Unsharp mask

Hβ

routines can be applied which enhance the lines and make them easier to identify (Fig. 9.2). I have found that a gentle sharpening routine and/or an unsharp mask produce a reasonable enhancement of features with minimal or no artifact introduction. Absorption and emission line identifications were carried out using tables from the *Handbook of Chemistry and Physics* (1998–9). Flux calibration, when required, was done according to published methods (Kannappan and Fabricant) and is indicated in Fig. 9.3. The method consists of using published absolute fluxes per unit wavelength for standard spectrometric stars (Fig. 9.3a) and normalizing of spectra of the same standard stars obtained with my optical arrangement (Fig. 9.3b–d) against these values. This general procedure is outlined in Fig. 9.3. Once the calibration has been carried out and a set of normalizing values obtained, these values can be used indefinitely as long your set-up remains unchanged. If simple identification of absorption or emission lines is all that is required, flux calibration can be eliminated. However, if more robust analysis of the data is to be carried out, flux calibration is essential. For example, in order to determine temperatures and electron densities in emission nebulae, flux calibration is required since a careful determination of line intensity ratios is needed for these type of calculations. This type of analysis will be demonstrated later in the chapter.

Figure 9.2. Example of processing steps involved in spectral analysis. The raw spectrum is only a few pixels wide and is initially processed as with any digital image with a dark frame subtraction. The image is then cropped into a 765 × 20 pixel swath, which allows it to be imported into SPECTRA software. The image is then expanded and calibrated with known emission lines from gas discharge tubes. Alternatively, the calibration can be carried out directly on the spectral features, provided that at least two lines can be identified. The expanded spectrum can be further processed to highlight subtle line features.

Figure 9.3. Flux calibration of spectra. **a** Digital flux calibration data for standard stars is downloaded from the appropriate database and is given as magnitudes per wavelength interval. This is converted to photons/cm²/sec/nm using the formula $F_\lambda = 5.5 \times 10^6 (1/\lambda) 10^{-0.4(mag)}$.

Wavelength A	Magnitude	photons/cm²/sec/nm
3604	6.643	33.6029888
3620	6.632	33.79512981
3636	6.625	33.86404363
3652	6.613	34.09038615
3668	6.609	34.06695858
3684	6.604	34.07556519
3700	6.564	35.20148342
3716	6.481	37.83441 16

a

Wavelength (A)	Raw counts/pixel	Photons/pixel	Photons/cm²/sec/nm	Std data	Correction factor
3704.88	2930	6739	0.057278121	35.2	614.5453035
3709.17	3135	7210.5	0.061285634	37.83	617.2735344
3713.45	3325	7647.5	0.064999915	37.83	582.000761
3717.73	3248	7470.4	0.063494654	37.83	595.7981929
3722.02	3422	7870.6	0.066896153	37.83	565.5033695
3726.3	3368	7746.4	0.065840515	41.88	636.0825054
3730.59	3375	7762.5	0.065977357	41.88	634.7632232
3734.87	3488	8022.4	0.068186377	41.88	614.198933
3739.15	3545	8153.5	0.069300661	41.88	604.3232379

b

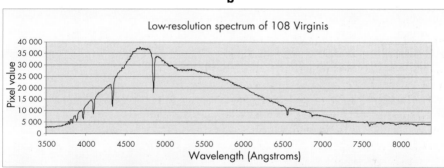

c

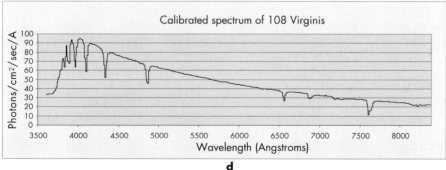

d

Figure 9.3. b The spectrum of the same star is obtained with your particular optical system and converted to photons/cm²/sec/nm. Division of the calibrated data by your results produces a correction factor, which is then applied at the appropriate wavelength intervals. **c** Spectrum of 108 Virginis before flux calibration and **d** after flux calibration

Results and Discussion

The low-resolution mode is useful for stellar classification and obtaining spectra of planetary nebula. In the high-resolution mode, many absorption lines of atoms, ions and simple molecules are visible. Figure 9.4 shows the higher-resolution spectra of stars from class B to M and luminosity class III and spans the region from Hβ to about 4100 Å. Several of the more prominent lines are labeled. As one can see, many lines are present, especially as one proceeds to cooler stars. Another way this can be examined is shown in Fig. 9.5, where the wavelength-calibrated results were imported into Microsoft Excel for further analysis. The region around Hγ (4300-4380 Å) has been expanded to show how the Hγ line profiles change with spectral type and how neutral iron (Fe) and chromium (Cr) lines are affected by stellar type. The line profile of relatively intense lines such as those for hydrogen contains information regarding the physics of the stellar atmosphere. For example, buried in the line profile are such information as pressure, density and temperature, some of which, with the appropriate mathematics, can be extracted. In addition, rotation of the star is also contained within the profile, which can also be extracted (Kitchin 1995).

Figure 9.4.
Classification of stars based on their spectra. Spectra were obtained in the high-resolution mode from Hβ to Hδ for stars from early B to M class. Note the increase followed by decrease in the hydrogen absorption lines as one proceeds from B to M type stars along with the general increase in the number of metal lines as the temperature decreases toward M class stars. Several different metal lines are identified along with the G band representing the diatomic molecule CH.

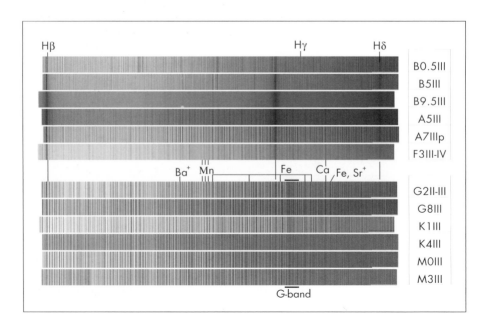

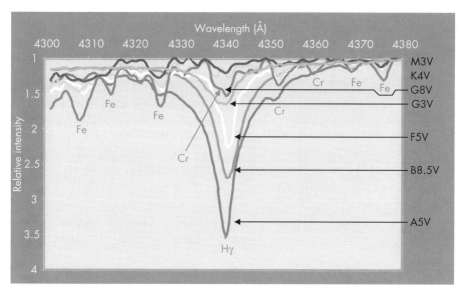

Figure 9.5. Graphical representation of the spectral sequence centered around the Hγ absorption line. Note how the intensity of the Hγ line, which has been normalized to 1 for all spectra at 4380 Å, increases to a maximum at type A stars and gradually falls off with cooler stellar types. Several lines for iron (Fe) and chromium (Cr) are also identified.

Figure 9.6 shows the emission line spectrum of NGC7009, the Saturn Nebula. The upper left corner shows an image of the nebula with the slit super-imposed over the nebula. The main body of the line spectrum was obtained in the low-resolution mode so that a majority of the emission features could be observed in a single picture. A higher-resolution spectrum is shown in the upper right centered at the Hα line. In high-resolution mode, three components are seen to make up this strong emission, Hα and two lines of N^+, which flank Hα. A graphical presentation of the line spectra is also shown. The usual types of species are apparent in the spectrum of a planetary nebula. The hydrogen and helium lines along with the prominent forbidden O^{++} lines, Ar^{++}, Ar^{+++}, N^{++} and S^+ round out the variety of ionic species present. For extended objects such as nebulae, one obtains a spectrum across the entire length of the slit. During the analysis of the results, the software allows you to select small subregions of the spectrum for analysis. As a result, one can profile the composition across the entire slit, effectively obtaining many spectra of the object at a single time, which is analyzed within the software. As discussed below, this would also allow for profiling temperature and density of the nebula across the slit.

In addition to identifying emission features, other physical features of a nebula can be determined, such as the temperature (T) and electron density (N_e). Theore-

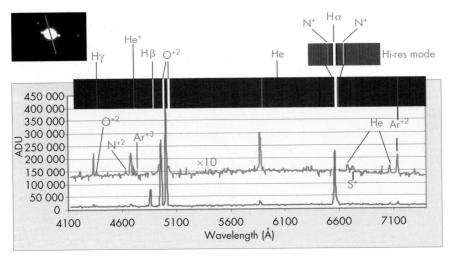

tically derived equations have been obtained which relate electron density and temperature to line ratio intensities:

$$(I_{4959} + I_{5007})/I_{4363} = [7.15/(1 + 0.00028N_e/T^{1/2})]10^{14300/T} \quad (1)$$

$$(I_{6548} + I_{6584})/I_{5755} = [8.50/(1 + 0.00290N_e/T^{1/2})]10^{10800/T} \quad (2)$$

Equation (1) relates the line intensity ratios for transition occurring at 4959, 5007 and 4363 Å for O^{++} while equation (2) relates the lines at 6548, 6584 and 5755 Å for N^+. If your spectral analysis is isolated to a small region of the nebula, you can assume that the electron density is the same in both equations and combine them to yield:

$$[(I_{6548} + I_{6584})/I_{5755}]7.15 \times 10^{14300/T} -$$
$$[(I_{4959} + I_{5007})/I_{4363}]0.82 \times 10^{10800/T}$$
$$= 0.9[(I_{6548} + I_{6584})/I_{5755}][(I_{4959} + I_{5007})/I_{4363}] \quad (3)$$

which is now independent of N_e. One can determine the intensities of the lines from your spectrum, assuming you have performed a flux calibration and use equation (3) to determine the temperature of the nebula (note that equation (3) cannot be solved explicitly for temperature but must be solved in an iterative fashion). As an example, for the Blue Snowball, NGC 7662, $(I_{6548} + I_{6584})/I_{5755} = 45$ and $(I_{4959} + I_{5007})/I_{4363} = 152$. This provides a temperature of 11 200° K (literature value

Figure 9.6. Spectrum of planetary nebula NGC 7009 (Saturn Nebula). In the upper left corner the positioning of the slit is indicated. The low-resolution spectrum is shown as both a graph and an emission line profile. The high-resolution spectrum is shown in the upper right centered around the Hα line showing the presence of N+ lines straddling the Hα line. Various other ionic and atomic species are identified. Many of these lines are forbidden, such as the intense 5007 Å line of O+2.

13 000° K (Lang 1980)). Taking this value for the temperature and using either equation 1 or 2 will give an electron density of 30,000/cm^3 (literature value 32,000/cm^3 (6)). This example portrays the gold mine of information contained within a spectrum. As was mentioned above, for an extended object the spectrum across the entire slit is obtained and line ratios as a function of distance from the central star, for example, can be obtained in a single spectrum. The potential is present to create two-dimensional profiles of temperature, density and composition for nebular type objects.

When the spectrometer is used in high-resolution mode, many absorption features can be observed in the spectra, particularly in cooler stars. Simple image processing techniques enhance these features making identification of features much easier. Figure 9.7 shows the high-resolution spectra of 78 Virginis, an Ap type star which exhibits enhanced quantities of europium (Eu), chromium (Cr) and strontium (Sr) in its outer atmosphere. A few of the many lines have been identified and are labeled in the spectrum. Iron, barium and chromium are dominant features. The identification of species responsible for the observed absorption lines remains somewhat of an art. *The Chemistry and Physics Handbook* lists over 21 000 lines between wavelengths of 3600 and 10 000 Å for all the elements. This represents on average about six lines per angstrom wavelength interval. Since the instrument is only at best able to resolve 1–2 Å in the higher-resolution mode, some criteria must be used to eliminate many of the

Figure 9.7. The optical spectrum of 78 Virginis, an Ap star of the europium–chromium–strontium class. These types of stars exhibit enhanced abundances of these heavy metals and the identification of absorption lines are labeled along with the identification of more common metal lines such as iron (Fe), sodium (Na) and potassium (K). The numbers indicate the wavelengths of selected absorption lines in angstroms.
Eu = europium,
Cr = chromium,
Sr = strontium,
Ba = barium.

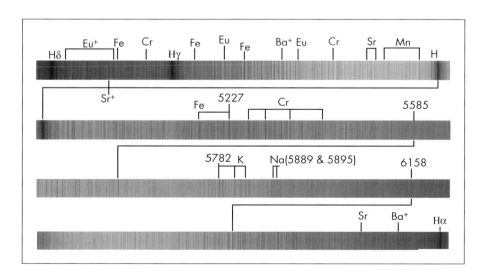

potential lines observed or, as is often the case, a line may represent a blend of two or more lines.

The presence and intensity of a feature, due to an atom or ionic species, is the result of many parameters such as natural abundance, probability of the absorption, temperature, density and pressure. For our purposes, the first three are the most important. Many potential features can be eliminated at the start simply because the natural abundance of an element is so low. For example, the Hβ absorption line at 4861.33 Å has a relative intensity of 80 compared to 3000 for protactinium at 4861.49 Å. Yet one would not typically assign this absorption line in a stellar spectrum to protactinium simply because the natural abundance of this element is 9 orders of magnitude lower than that of hydrogen. In addition to this abundance factor, other absorption lines for hydrogen are present where they should be whereas protactinium lines are not. In order to be certain of an assignment of a line, it is important to find several lines for a particular species. Thus, as you can see in Fig. 9.7, the lines for europium, chromium and strontium have many lines which are apparent for each species. This makes you more confident that your assignment is real. In addition, for all the spectra shown in this chapter, the spectra obtained were compared directly with those obtained in the astronomical literature to be certain of the assignments. My initial goals in using this instrument were to establish how far the instrument could be pushed in giving spectra *and* in the resulting identification of features.

Figure 9.8 illustrates just how far this instrument is capable of being pushed to give useful data. R Andromeda is an S3.5e–S8.8e (M7e) Mira type variable star with a period of 409 days. Stars of this

Figure 9.8. The spectra of R And, an S3.5e–S8.8e (M7e) Mira type variable star with a period of 409 days. The upper spectrum runs from 4100 Å to 4900 Å while the bottom spectrum runs from 5800 to 6600 Å. Note the presence of emission lines for Hα, Hβ, Hγ and Hδ lines along with the presence of the unstable element technetium. In addition to enhanced amounts of ^{13}C as seen in diatomic carbon lines, zirconium oxide molecular bands are also observed.

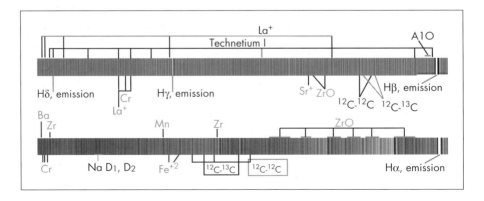

spectral class often exhibit absorption lines due to the unstable element technetium. This element is not found naturally in the solar system because its longest lived isotope, ^{97}Te, has a half-life of only 2.6×10^6 years and as result all technetium endogenous to the formation of the solar system has long since decayed. Yet multiple lines of technetium can be detected in the atmosphere of this star and others of S and C types (Jaschek and Jaschek 1987). Apparently, neutron capture is proceeding deep within stars of this type followed by a dredging mechanism, which brings to the surface nuclear processed material. Note also the presence of zirconium and zirconium oxide molecules, which give rise to extensive banded structure in the spectrum in the red region. In this particular S type star, emission lines are observed superimposed upon the more usual hydrogen absorption lines.

The other very interesting aspect of these type stars is the fact that they often contain abnormal amounts of ^{13}C compared to ^{12}C. The normal solar system ratio of ^{12}C to ^{13}C is 80, but in many of these type stars this ratio can approach 4. Normally, the detection of isotopes of atoms or ions cannot be done easily because the lines are extremely close together and normal line broadening effects cause them to overlap. However, this is not the case with molecules where rotation and vibration transitions are much more sensitive to the isotope composition. These types of transitions are normally outside the optical range but electronic transitions can couple with vibration transitions giving rise to a blanketing effect of the multitude of lines that result and these lines are often observable in the optical region. In cooler type S and C type stars diatomic carbon forms and a clear separation of absorption lines occurs between diatomic ^{12}C–^{12}C and ^{12}C–^{13}C as is indicated on the spectrum in Fig. 9.8.

Figure 9.9 shows the spectra of a variety of type C and R stars containing varying ratios of ^{12}C to ^{13}C. Some of these stars are noted for their large quantities of carbon as observed with diatomic carbon. The upper pair of spectra represents stars with ratios in the area of 10. All three possible combinations of diatomic C-C are observed, ^{12}C–^{12}C, ^{12}C–^{13}C and ^{13}C–^{13}C. The middle pair of spectra represents stars with higher ratios such that ^{13}C–^{13}C is no longer clearly observed and finally for stars approaching solar system ratios, only ^{12}C–^{12}C is seen (lower spectrum). The blanketing effect of diatomic carbon is shown along with the identification of several metal lines.

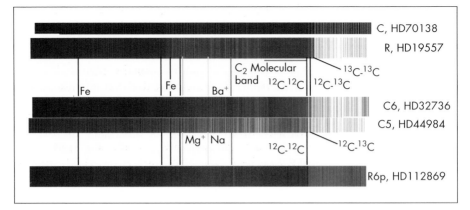

In addition, with careful wavelength calibration one can measure the Doppler shift of absorption and emission lines to determine velocities of approach or recession of objects along with rotation velocities of stars and planets. I have been able to successfully observe and measure the rotation of Saturn by aligning the slit along the rings of the planet. In exposures of only a few seconds one is able to obtain a spectrum of the entire disk and ring system of Saturn. The eastern and western region of the image can be isolated using SPECTRA software and each of the spectra calibrated against the Hα line. When the final two images are aligned they show the clear shift in the lines which occurs due to Saturn's rotation from which can be calculated the rotation velocity. In a similar manner, the velocity of approach of M31 has been determined by acquiring the spectrum of the nucleus of the galaxy. In this case only the narrow 18-μm slit was used and as a result a relatively long 60-minute exposure was necessary. I have not tried using the wider slit but according to the instrument specifications, a gain of 2 magnitudes is possible, with of course a loss of resolution, which would occur with the wider slit, but one should be able to determine the center of lines with the software provided.

Figure 9.9. Spectra of type C and R stars containing varying ratios of ^{12}C to ^{13}C. These stars are noted for their large quantities of carbon as observed with diatomic carbon. The solar system value for this ratio is 80. The upper pair of spectra represents stars with ratios in the area of 10. All three possible combinations of C–C are observed: $^{12}C-^{12}C$, $^{12}C-^{13}C$ and $^{13}C-^{13}C$. The middle pair represents stars with higher ratios such that $^{13}C-^{13}C$ is no longer observed and finally for stars approaching solar system ratios, only $^{12}C-^{12}C$ is seen (lower spectrum). The blanketing effect of diatomic carbon is shown along with the identification of several metal lines.

Future Directions

The future is certainly bright for amateur spectroscopy. As far as the SGS is concerned, future versions will offer the option of having a higher dispersive grating (1800 lines/mm). This will improve in the identification of lines by spreading the spectrum out by a factor of three compared to the current high-resolution grating. In addition, user friendly and versatile software for the amateur spectroscopist would be most welcome. While IRAF is the current standard for this type work among professionals, its requirement of a Unix or Linux operating system, along with what I understand to be a difficult package to learn at best will make this system little used by the budding spectroscopist. To this end, perhaps the spectroscopy module currently being developed by Axiom to be part of the MIRA software package will fill the void.

Conclusions

The Santa Barbara Instrument Group Spectrometer represents a quantum leap forward for the amateur interested in the fertile area of spectroscopy. Even with a relatively small telescope, this instrument coupled to sensitive CCD cameras and utilizing the self-guiding feature of the ST-7/8 camera allows one to reach unprecedented magnitudes and carry out spectral analysis only dreamed of by the amateur a few years ago. Even after using the instrument for a year, I remain astounded by the fact that an amateur with only relatively modest equipment from his own backyard can detect technetium, many dozens of other elements, simple molecules and carbon isotopes in stars or nebulae hundreds of light years away.

Reference

Handbook of Chemistry and Physics, 79th edition, 1998–9, section 10-1 to 10-88.

Jaschek, C and Jaschek, M (1987) *The Classification of Stars*. Cambridge University Press.

Kannappan, S and Fabricant, D (2000) "Getting the Most from a CCD Spectrograph." *Sky and Telescope*, 100(1), pp. 125–32.

Kitchin, CR (1995) *Optical Astronomical Spectroscopy*. Institute of Physics Publishing.

Lang, KR (1980) *Astrophysical Formulas*, 2nd edn. Springer-Verlag.

MaxIm, Cyanogen Productions Inc., www.cyanogen.com/ SBIG, www.sbig.com

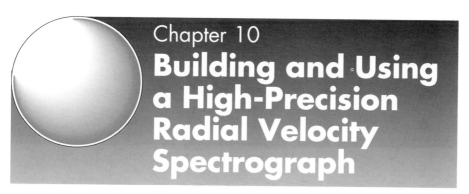

Chapter 10

Building and Using a High-Precision Radial Velocity Spectrograph

Tom Kaye

This chapter covers the construction of a high-performance, stabilized, fiber optic spectrograph capable of precision radial velocity measurements down to about 200 meters per second. This level of velocity detection is better than a high percentage of professional systems. Building a spectrograph can be a rewarding project for amateurs with a desire to do "real science" and looking to expand their knowledge base. The construction concepts are simple and straightforward; maximizing the performance of the system requires software and hardware finesse. This spectrograph can detect shifts of 3000 Å or one hundred-thousandth of an inch. It was the first amateur system to detect an extrasolar planet using radial velocity. Some may find the challenge intriguing as the author has and this chapter should act as a guide.

Layout

The optical layout of the system is shown in Fig. 10.1. It is a typical Czerny–Turner design with separate collimator and camera mirrors that can be swapped out to adjust resolution. A secondary mirror reflects the camera beam into the CCD, effectively limiting stray light entering the detector. Multiple fibers link the spectrograph to the telescope and reference lamp. The fiber optic cable allows for a very robust and stable

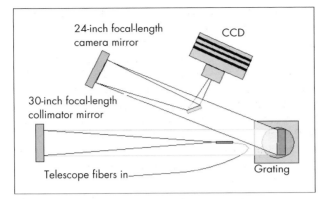

24-inch focal-length
camera mirror

CCD

30-inch focal-length
collimator mirror

Telescope fibers in

Grating

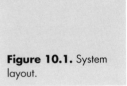

Figure 10.1. System layout.

bench mounted system that does not have to be mounted on the telescope.

Performance

All spectrographs are rated by their resolution. Many amateurs think this is defined by the spectrum's dispersion per pixel but that is only part of the formula. To calculate resolution, first determine the FWHM (full width at half maximum) of the fiber or slit projected on the CCD. Imagine that the system did not spread the light into a spectrum but just imaged the slit on the detector. For all practical purposes you cannot see something smaller than the slit and this is your limiting factor in resolution. If each pixel covers one angstrom of spectrum and the slit image is three pixels wide, then logically you have a 3 Å slit. In practice the absorption lines will have a FWHM that is close to your slit width. If the middle of the spectrum on the CCD is, say, yellow, then to find your resolution, you divide the 6000 Å (for the yellow) by three (the slit width in angstroms) to get a resolution of $R = 2000$. Resolution below 10 000 is generally considered low, 10–25 000 is in the medium range and above 30 000 is considered high resolution in the professional world. Most spectrographs that have historically been available to the amateur have had R in the hundreds (rainbow gratings) or a few thousand for some commercial models. The system described below has a dispersion of about one tenth of an angstrom per pixel and goes out to $R = 15\,000$. This puts it squarely in the medium-resolution range, far above the average amateur system.

The spectral images generally cover less than 100 Å across the face of the CCD. As an example you can see lines in between the sodium doublet.

The resolution is necessary so you can "see" the small shifts in the spectrum from red shifts and blue shifts but also requires system stability to make really fine measurements. Temperature changes as well as barometric pressure shifts can introduce spurious shifts in the spectrum, swamping your readings. The design incorporates heating and cooling controls to hold the temperature within one degree for extended periods. In addition to temperature control, mechanical stability is also of prime importance. The system's optical base and mirror mounts are built almost exclusively out of granite. Granite offers great thermal and mechanical stability at bargain prices.

The last part of the performance envelope is determined by the size of the telescope. The more resolution you have, the fewer photons per pixel there are to accumulate in a given time. This type of resolution is almost exclusively for brighter stars; galaxies are completely out of the question. In practice we have successfully used a 16-inch scope on 4.5 magnitude stars with a 45 minute exposure. Going deeper will require a bigger scope or lower resolution.

Fabrication

The major components to be fabricated are as follows:

- thermal housing, consisting of insulating foam board;
- optical platform, made from granite counter top;
- optical component mounts, from granite tiles and some machined parts;
- fiber cable, a combination of glass fibers and bicycle brake cable with machined end;
- temperature control system including a pad heater and a water cooler chiller.

Additional items needed:

- parabolic mirrors for the camera and collimator;
- grating for spectrum dispersion;
- rotating base for grating;
- secondary mirror in front of CCD;

- CCD, AP7 or ST7 equivalent that should be water cooled.

First we'll start with the housing: it is the easiest and goes together rapidly. Most construction supply stores sell foam insulation in 4 × 8 foot sheets at very reasonable prices. Purchase several of these along with strong glue and duct tape. Razor blades will cut the foam very efficiently. Construct a box with a final dimension of four foot long by three feet wide and two feet high. This will be double walled as shown in Fig. 10.2 with an air gap between the inner and outer walls. A tongue and groove joint is incorporated between the base and top by making the air gap the same size as a sheet of foam. A small section of tongue is left open and replaced with

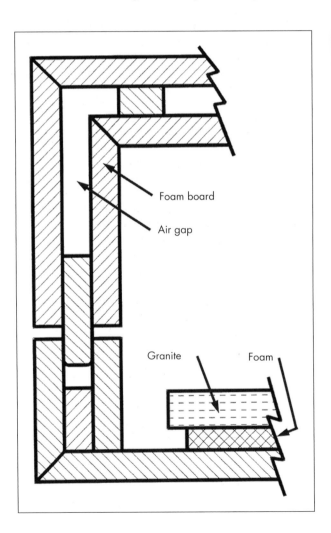

Foam board

Air gap

Granite

Foam

Figure 10.2.
Cross-section of double walled thermal enclosure.

flexible foam to allow routing of fibers and electrical cables. Start by assembling the inner box and gluing the joints while using the duct tape to hold the sides together. The bottom is only one sheet thick because it has to hold the weight of the spectrograph. Proceed with the outer shell and don't forget to reinforce the sides and top in a few places by connecting the inner and outer layers with small blocks of foam. Some foam board comes coated with thin aluminum foil and can be used to improve the external appearance. Finally, cover the inside with a suitable light-absorbing material such as black felt. This should produce a very light but rigid shell that can be removed by one person. The entire housing should rest on a concrete basement floor or rigid steel framework. Do *not* expect good stability from wooden work benches.

The optical platform is made from polished granite commonly used for expensive counter tops. Good success has been had with Cambrian Black Granite and should be available in 1.25-inch thick slabs. Try to find a piece that will fill the inside of the enclosure with a several inches of clearance on each side and have it cut to size at the dealer. This will act as the optical base that will sit on an inch-thick piece of urethane foam the size of the slab. After all the components are assembled, they can be positioned on the slab for testing and determining their final placement. At this point, once your final placements have been determined, the slab can be marked for drilling the mounting holes. A carbide bit and hand drill with coolant can be used to penetrate the granite. In order to thread bolts into the slab, threaded brass inserts [1] (see the list of suppliers at the end of the chapter) commonly used for plastics are inserted from the back side. These inserts use a slightly larger drill for their mounting holes so the slab is turned upside down and drilled part way through to accommodate the insert. The inserts are wedged into place with the mounting bolts.

Figure 10.3 shows the clamp arms used to hold down the components while still allowing some range of adjustment. The clamps can be simple pieces of aluminum with holes drilled for the mounting bolts. 1/4-20 threaded bolts work well here.

Mirrors for the system can be purchased locally or from suppliers such as Edmund Scientific [2]. The collimator mirror can be either 4-inch or 6-inch diameter parabolic with a focal length of 30 inches. The camera mirror should also be parabolic with a focal

Figure 10.3. Clamp arm and focusing mechanism.

length of 24 inches. It is well worth the efficiency increase to get multicoated mirrors that only lose about 1% of the reflected light. Different focal lengths can be substituted with a corresponding change in resolution. Spherical mirrors can be used if nothing else is available.

Ultrastable mirror mounts are made from 1 × 1 foot granite tile $\frac{1}{4}$-inch thick purchased by the box. Cambrian Black is again the granite of choice. A diamond tile saw is used to cut strips of granite that are epoxied together as shown in Fig. 10.4. The mirror sits directly on the granite at the lower two points. The third mounting point on top is a nylon tipped set screw threaded into another brass bushing epoxied into a hole in the granite frame. The set screw allows for vertical positioning of the mirror and the whole mount is rotated left and right for alignment.

The entire camera mirror mount is moved to focus the image on the CCD. A square aluminum bar is drilled and tapped in the middle to accept a long threaded bolt (Fig. 10.3). The threaded tip of the bolt is rounded and the aluminum bar can be clamped into position in front of the mirror mount. The focusing bolt's rounded tip pushes against the mirror's granite base in the center. An aluminum right angle glued to the granite base allows the mirror mount to slide without rotating. Focusing is accomplished by trial and error once the system is at a stable temperature. Since the whole mechanism is temperature controlled the system, once adjusted, will stay in focus, so sophisticated mechanisms are not required.

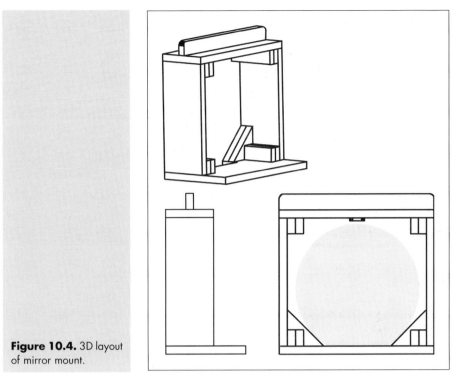

Figure 10.4. 3D layout of mirror mount.

The grating is the heart of the system and great care should be taken not to touch or even breathe on the reflective surface. The raw aluminum coating can get acid etched by the moisture in your breath. A minimum 50 mm square 1800 line grating can be purchased from supply houses for a reasonable price. Ruled gratings are generally more efficient than holographic and should be blazed at 500 nanometers. For those with a larger budget we highly recommend a 4 × 4-inch grating from a major supplier [3]. The larger grating loses less light from the collimator and greatly reduces the exposure time. The several thousand dollar price tag may take some getting used to.

The grating mount should be fully adjustable and rotate to place the correct section of spectrum on the CCD. A simple but effective mount can be made from steel angle iron that is silicone glued at three spots to the back of the grating. Make sure to leave some gap so you can cut the silicone later if you want to remove the grating. Also check to make sure the grating is reflecting the right way for maximum brightness; an arrow should indicate the blaze direction. The base is another steel U

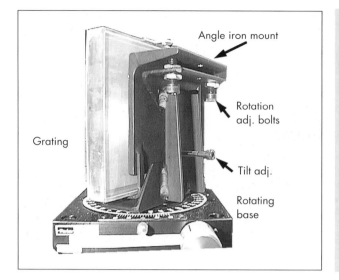

Angle iron mount

Rotation adj. bolts

Tilt adj.

Rotating base

Grating

Figure 10.5. Grating mount.

channel that is drilled and tapped in three places, see Fig. 10.5. Two bolts with pointed tips facing upward on the top of the U channel interface with small drilled holes on the grating mount. Adjusting the bolts up and down will rotate the grating on axis to align the spectrum with the rows of CCD pixels. The third hole accepts a bolt that pushes the back of the grating mount to adjust the reflected beam vertically. This mount allows the grating to be easily lifted off for safe keeping while working on the system. A safety wire is recommended in case the grating is knocked off the mount. The whole assembly is mounted on a rotating stage [2] to select the frequency range of interest.

The secondary mirror can be purchased off the shelf [2] and glued to a small steel shaft (Fig. 10.6). An aluminum or steel block is drilled for the shaft and a set screw tapped in the side to lock the shaft in position. The base is clamped in position with a mounting screw and clamp arm. The mirror will adjust vertically and axially about the shaft and after manual positioning is locked in position with the set screw.

The CCD is the business end of the system and the assumption is made that the reader is completely familiar with its use and operation. It is highly recommended that the user considers a thinned, back-illuminated camera such as the Apogee AP7 used in the author's system. Although more expensive than competing front-illuminated units, the fact that the AP7

Figure 10.6. Picture of secondary and mount.

has almost double the quantum efficiency is equal to doubling the effective aperture of your telescope. The camera should be water cooled so the excess heat is removed entirely from the sealed enclosure. While in use, adjust the TE cooler to run constantly, as the chip will move when the cooler cycles. A shutter is highly recommended to simplify dark frames and exposures. For best results choose a camera with low read noise and little dark current. While not a large factor in producing astrophotos, noise is a big factor in precision measurements.

The CCD mount should be machined out of $\frac{1}{2}$-inch steel or aluminum. It is basically a large "L" bracket with mounting holes for the CCD and base bolts. Side plates are bolted on to increase the rigidity and reduce thermal movement. Figure 10.7 shows an aluminum mounting plate. The trick in using this mount is to insulate the CCD from the mount by using Teflon spacers between the CCD and mount. The water cooling chills the camera housing and the heat is drawn out of the mount and eventually the granite, causing a cold spot at that end of the spectrograph.

The cooling system is a chiller from a drinking fountain [4]. They are inexpensive but require an additional circulation pump to move the water. Inside the spectrograph you can fabricate a thin aluminum box tube running the length of the spectrograph that will serve as a cooling plate. It can be mounted at some convenient point along the rear of the spectrograph but should not touch the granite directly. The water flows through the cooling plate, to the CCD and back to the

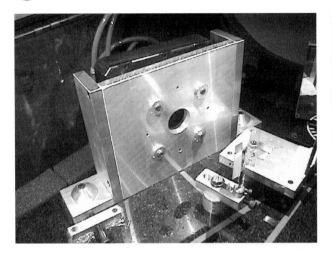

Figure 10.7. CCD mount.

chiller and pump. Isopropyl alcohol is usually mixed about 10% with the water to prevent mold build-up. This system runs full time to maintain the system temperature which is usually about 20° C. If the enclosure is in a temperature controlled room, the cooling system may settle into a very stable temperature range. You are looking to maintain the enclosure within about one degree. If that is not possible then you can add a heating system to bring the temperature up a couple of degrees from the coldest point. This can be accomplished with a small electric heating pad in the top of the enclosure controlled by an industrial temperature controller [5]. Watch out for condensation in the system; this may require adjustment of the temperature settings.

Because no mechanical system is perfectly stable and barometric pressure changes can affect the data, it is a requirement to have a reference spectrum. Low-priced educational hollow cathode lamps [2] with various gases are available. These are useful for low-resolution work and to figure out where you are on the spectrum. With continuous use they become contaminated by the electrodes and have a limited life. A low-pressure sodium lamp is very convenient for determining your dispersion with the sodium doublet.

For serious radial velocity work the reference spectrum needs lots of emission lines for good measurements. The thorium–argon lamp [6] is the most widely used reference source for professional spectrographs. It produces thousands of lines across the entire spectrum and is guaranteed for 5000 hours of

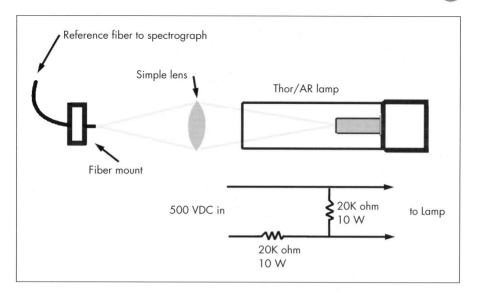

Figure 10.8.
Thorium/argon lamp layout and power supply.

operation. In practice we have not heard of one going bad. There are professional power supplies [7] that, while expensive, offer precise control over the voltage. The author's system used one of these supplies but an inexpensive alternative has turned up and a schematic is shown in Fig. 10.8. This wiring diagram comes from Ken Nordsieck of the University of Wisconsin/Madison. It is possible these lamps have "flows" which register as shifts in the spectrums and it is not known if this less expensive power supply will cause problems. Use at your own risk.

The reference lamp is mounted outside the enclosure on its own platform. A simple lens focuses the cathode on the reference fiber which transmits the light into the spectrograph. A typical mounting arrangement is shown in Fig. 10.8. This is also a convenient place to mount some sort of flat field lamp that can put white light down the reference fibers. The entire unit should be shielded from stray light.

Fiber Optic Cable

The fabricator should plan on spending the time to properly construct and polish the fiber optic bundle. The more precisely you make this part the easier time you will have on every observing run. The quality of the polish on the fiber ends will greatly affect the efficiency of the system. The fibers are glass and, while flexible, will snap if overstressed. Securing the fibers and making sure nothing comes loose or kinks will insure a long life for your bundle.

Most professional systems use fibers from Polymicro [8]. The system described here uses the 100 micron, step index, low OH fibers part # FVP 100 120 140. These fibers come with a brown plastic coating that is relatively tough but they should not be abused or bent too far. There are seven fibers from the telescope to the spectrograph and two others to the reference lamp, so order a sufficient length.

The tightly packed seven-fiber telescope cluster allows for some star misalignment but it will still show up on one of the fibers. Additionally all seven fibers can be tested for efficiency and the best ones used preferentially. Long runs are generally not a big problem and most of the losses occur at the ends due to polishing, etc. Another factor to be aware of is focal ratio degradation; high-quality fibers do a reasonable job of preserving the angle of the input light cone in the output. Put another way, if you have an f/10 input beam you are likely to get about an f/6–f/7 fiber output beam. This is important because unless you have an F/6 collimating mirror you will lose light. If you have a wider field telescope which puts an F/6 beam in the fibers you are likely to get F/4 out which will cause more light loss at the collimator. If one constructs this system with 4-inch mirrors there will be considerable light loss but with the benefit of cheaper optics and grating. By carefully minimizing the curves in the fiber bundle and using a high focal ratio telescope, you will minimize focal ratio degradation and get the best efficiency.

The best and cheapest armored cable is typical bicycle brake cable [9]. It is vinyl coated and Teflon lined with a spiral wrapped spring steel coil that limits the bend radius. It can be purchased in 75 foot rolls directly from the manufacturer. This cable is ready to go as is with only the addition of the telescope fruel.

The telescope interface has several parts: a fruel, eyepiece adaptor and fruel tip (Fig. 10.9). The fruel is the fiber's terminating end and is plugged into the eyepiece adaptor. A standard eyepiece is used to focus the scope. It is then replaced with the eyepiece adaptor which is pinned, so it locates rotationally in the same place every time. The fruel is then inserted into the adaptor and it too locates with a rotation pin. A set screw in the shoulder of the fruel adjusts in and out so

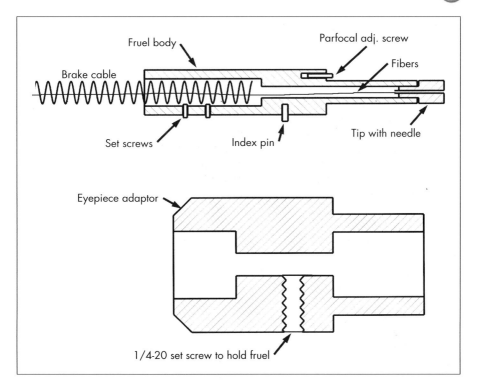

Figure 10.9. Fruel diagram.

the fiber ends are parfocal with the eyepiece. A set screw in the side of the eyepiece adaptor locks the fruel in place. These parts should be precision machined out of aluminum or brass. Minimum clearances will allow for better repositioning if the adaptor has to be removed for refocusing, etc. Since you are likely to use an off-axis guider the more repeatable the fruel position is, the easier time you will have acquiring the star of interest.

Laying out the fibers and threading them into the armored cable is a point of high frustration for many spectrograph engineers. We very highly recommend the following procedure. Find a long hallway twice the length of the proposed cable. At one end, lay the cable down one side and run the same length stiff wire down the cable until it comes out both ends. Next lay out the fibers at the other end of the hallway in a straight line one at a time by carefully rolling them off the spool by hand. Now use super-glue to adhere the fibers to the end of the wire. While one person pulls the wire slowly through the cable, another lets the fibers run through his hands to feel for any hang-ups or sudden tension. When finished, there should be a few feet of fiber hanging out both ends of the cable. To protect the exposed fiber, slide a length of pvc tube over the fibers and armored cable, then tape in place. This will allow transport to the polishing bench.

The fibers are now ready to be glued into the fruel assembly. First thread the exposed fibers through the fruel body and set screw the body to the cable using Superglue on the set screws. Next cut off the excess fibers leaving about six inches to work with. Thread the seven fibers through a 3/4 inch length of hypodermic needle [10] that has both ends polished to remove any burrs. The needle should be sized so that the fibers fit loosely but self-arrange into a circle of six with one in the middle. This is known as a 6–1 wrap. Use the slowest drying epoxy available, preferably several hours, *not* the five-minute variety. Fast drying epoxies do not fully harden at the scale of the fibers and will make polishing impossible. Saturate the fibers and pull them backwards until the needle is completely filled with epoxy and fibers. Wipe the outside of the needle clean and let dry over the next 24 hours for maximum stiffness.

The fruel tip holds the fiber/needle securely in place. Assemble the needle and tip so there is some needle sticking out of the tip and use epoxy or Superglue to secure it in place. Lay out the fiber cable in a straight line and while an assistant gently pulls on the fibers, seat and press fit the fruel tip into the fruel body. The fiber is now safely in the cable and this end can be polished.

Fiber polishing is somewhat of a black art. Different things seem to work for different people so if one method is not successful try something else. A plastic puck with a hole in the center is used to position the fruel assembly perpendicular to the sanding base which is usually a solid plate of aluminum. Start by clipping off the extra fibers and use progressively finer sand paper to sand off the excess needle down to the fruel tip. Use finger pressure on the fruel to judge the material removal. As you get closer to the tip use the finest sandpaper available, usually 1200 grit wetsand. A 75–100× power microscope is used to view the fiber ends at all stages of polishing. Once the fiber ends are in plane with the fruel tip, circular motions are used to get as smooth a finish on the fiber ends as possible. It is fine to polish the fruel tip. A good final polish has been obtained with a liquid brass polish called Brasso on plain manila folder paper. Change the paper often and use the microscope to look for a smooth, flat, glass-like surface on the fiber ends. There should be no pits and the fibers should look like deep black wells when polished properly.

At the spectrograph end the fibers are left loose out the end of the cable. This is done so coiling up the cable will allow the fibers to move in and out at will. Great care must be taken so as not to damage the fibers in the next steps. Clamp the cable down on the work surface and one at a time illuminate the ends of the loose fibers with a strong source. Looking under the microscope at the fruel end, note which fiber in the 6–1 wrap is lit up. The idea is to arrange the fibers in the spectrograph in a single row that corresponds to some pattern in the 6–1 wrap. This tells you which way to move the scope depending on which fiber is illuminated. Once you have the fibers organized, tape them close together along with two more fibers for the reference lamp at each end. Your row of nine fibers is now epoxied to the side of a steel vane which will be used for the mount. The fibers should be in a straight row and touching each other. Once the glue has dried, clip off the excess fibers and you are ready to do the same polishing routine on this end. This time there should be the few feet of extra fiber between the mounting vane and the cable end so be careful. The vane is held against the side of a square plastic block to facilitate holding it at 90° to the work surface. Again polish to a mirror finish. The steel vane is bolted to a base mount that is clamped to the granite slab.

At this point all of the major components should be fabricated and working. It will be assumed that anyone with enough skill to produce these parts can assemble and optically align the system so far described. All components are on-axis to minimize design and alignment problems. Specific dimensions were left out with the concept that each system will use whatever parts are most available. Make sure to mount your optics high enough to clear the mounting bases and clamp arms.

Some tips on set-up: use a bright light down the fibers so the output beam can be positioned on the collimator mirror. The collimated beam should then be bright enough to align on the grating. The collimated spectrum will just barely be visible with a white card held in front of the camera mirror to get it centered. The spectrum in front of the secondary can easily be seen with the white card. Looking into the open shutter of the CCD, the spectrum will be visible, reflecting off the surface of the CCD when it is in position. The reflected spectrum on the CCD face should indicate what color band the spectrum is in. The reference lamp can be sent down all the fibers for a first light exposure. You should see nine rows of dots across the entire field.

Once sealed, the system will take about a day to temperature stabilize. Then the cover can be removed for short periods to adjust the final focus of the system. An electronic thermometer with min and max recording is a convenient way to monitor temperature stability.

You now have a fine precision instrument that should be capable of detecting stellar movement in our galaxy. It can be used for the study of elements as well as radial velocity. The hardware side is finished but the software is yet to come. Additional information is available at www. spectrashift.com <http://www.spectrashift.com>

Precision Radial Velocity Measurements

Commonly known as red shift and blue shift, radial velocities are caused by movement toward or away from a light source. Similar to the familiar shift in tone of a car horn as it goes by, the shift in the frequency of light brings a new view of the universe to astronomers. Identification of these shifts is accomplished by precise measurements of absorption lines which are dark bands in the star's spectrum caused by absorption from different elements. The measurement of these shifts, which can be smaller than one hundred-thousandth of an inch, is one of the last untouched areas for amateur astronomers. New capabilities have evolved in the past decade and amateurs have demonstrated remarkable capabilities such as discovering comets, supernovae and photometric measurements. Amateurs now compete favorably in astrophotography, photometry and astrometry. Spectroscopy is the only area where the amateur astrophysicist has lagged behind.

Spectroscopy is an appealing pursuit for astronomers living in light-polluted cities because much work can be done "in between" the bands of light pollution so the sky is effectively black. Since more amateur astronomers suffer from light pollution than not, spectroscopy should be ready for a renaissance. This becomes even more interesting when one realizes that everything in the universe is moving. Every star has a velocity, spectroscopic binaries are easily detected and things like sunspots can be detected in the unresolved point source of a star. This chapter outlines the process of measuring spectral shifts with high precision.

The techniques described below were part of the detection methods used by amateurs on the Spectrashift Team [11] to detect the extrasolar planet around the star Tau Boo in February of 2000. This software processing along with a stabilized spectrograph is capable of measuring relative speeds down to two hundred meters per second. This is well beyond most professional spectrographs and opens many possibilities for motivated amateurs. This level of precision was not thought possible at the amateur level until now. Spectroscopy requires an increase in exposure times because the photons are spread out over more pixels. This is an advantage for amateurs who usually have more time to devote and long-term monitoring of objects is likely to be quite profitable.

The intent here is to outline all the steps involved as a road map for implementation. It assumes that you already have a high-quality spectrum and reference spectrum in two stripes from left to right in a CCD FITS image. You should also be familiar with spectrum plots, absorption and emission lines. The techniques described here are for a fiber fed spectrograph with multiple fibers, with some fibers linked to reference lamps.

Basics

Spectral stripes on a CCD image are converted to an intensity graph that will plot the absorption features as peaks, as shown in Fig.10.10. Each peak has a particular shape known as a Gaussian distribution or bell curve. The center of each peak defines the position of that particular element's absorption frequency and only changes due to velocity. The simplest way to see velocity is to compare the spectra from two different stars with very different speeds. The formula to covert angstrom shift to velocity is

$$(\text{ang shift} \times \text{central wavelength of spectrum})$$
$$\times\ 300\,000 = \text{km/s}.$$

In order to visually see a two-pixel difference in peaks for two bright local stars with a difference of 100 kilo-

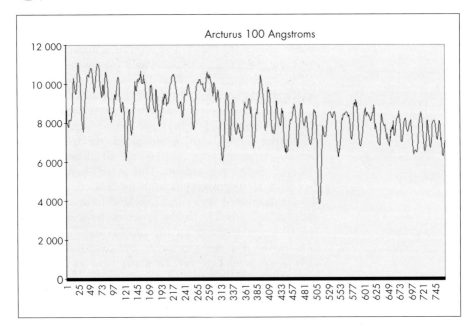

Figure 10.10. Intensity plot of Arcturus spectrum, 100 Å in the blue.

meters per second (km/s) one would need a dispersion of at least 1 angstrom per pixel with an appropriately small peak width. Figure 10.11 shows stars Zaurak and Menkar with each star's spectrum plotted against a reference Hβ spectrum below. There is an obvious difference in the absorption line in each star corresponding to the velocity difference. This is the most basic way to detect radial velocity.

There are various software programs on the market that are capable of finding the center of a single peak to subpixel resolution. With this method it is possible to see velocity shifts down to about 6 km/s. MaximDL [12] is working on modifying the star centroid function to work on spectra. This would allow a convenient way to look for larger spectrum shifts.

Fourier Cross-Correlation

Fourier cross-correlation (FCC) is by far the most widely used professional technique for measuring radial velocity. The advantage of this type of mathematical processing is that it uses the entire spectrum when making comparisons. It is orders of magnitude more

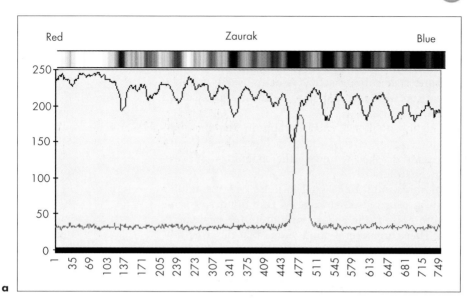

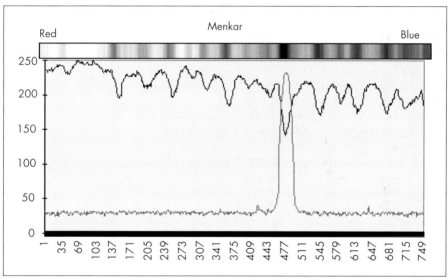

Figure 10.11. Zaurak and Menkar showing visual shift compared to hydrogen beta reference.

sensitive than other methods. Modern software allows the user to avoid understanding the underlying mathematics. FCC makes it possible to detect shifts as little as one hundredth of a pixel or 200 m/s but is not currently available in amateur software. There is, however, free professional software that accomplishes this feat with amazing simplicity.

High-precision FCC described here detects small velocity changes in the same star over time. It uses the star's first exposure to compare with all subsequent exposures. In this way any small changes in the star's speed can be detected. This technique, while sensitive, does not tell you the star's actual velocity, just the change from whatever velocity it was in the first exposure. Determining actual velocity usually involves comparison with a similar star of known velocity albeit at lower resolution.

A highly stabilized spectrograph in the medium-resolution range is required to achieve these results. The reader is asked to review the chapter on spectrograph construction.

Exposure times and signal-to-noise become important in measuring radial velocity. One hundred to one signal to noise is the general minimum for this type of work. There are several good works that describe how to calculate your signal to noise but a good rule of thumb is 10 000 photons per pixel. This is not pixel counts as many cameras require 2–3 photons per pixel count.

Software

The two software programs described below are MaximDL [12] and IRAF. Maxim is a consumer program and IRAF is a professional package available through the National Optical Astronomical Observatories (NOAO) [13].

The author highly recommends MaximDL software to control the spectrograph's CCD. This software has a nicely developed calibration routine that allows for averaging multiple dark and flat frames to reduce noise. It also allows for auto-exposure and incrementing of file names for unattended exposures. This is very helpful when running hundreds of exposures for stability tests on the system. MaximDL is also part of the ASCOM initiative which links diverse software together via scripting languages for automated control. This software gets the image out of the camera and into a FITS file.

NOAO spends US tax dollars to build telescopes and equipment. They also developed and distribute a software package called IRAF that is the most widely used professional package in the United States. It has over 1 000 000 lines of code and is free for downloading. A better idea is to buy the CD-ROM for less than $100.

IRAF is a UNIX program but recent advances in PC technology allow it to run on home computers. Most IRAF implementations run under the Red Hat Linux operating system. Implementing this operating system is beyond the scope of this chapter and is best left to a professional. Your consultant should be able to get you set up on Linux and get IRAF installed. He should get your Xgterm and Ximtool windows set up so you are ready to run. Typing **cl** at the command prompt will start the program. For the rest of this chapter computer commands are in bold lettering; commands are case-sensitive.

Navigating IRAF

IRAF is a collection of "packages" that have to be loaded to work. IRAF sprawls in that there are hundreds of programs that do almost everything in astrophysics. The most important packages are **noao**, **imred**, **kpnocoude**, **rv** and **ccdproc**. In these packages there are individual programs that will do the data reductions. They are **dofibers**, **fxcor**, **mkflat**, **mkdark**, and FITS conversions **rfit** and **wfit**. The very first step is reading *The Beginner's Guide to IRAF* in the help files. You can download this file from the NOAO website when you order the CD-ROM. If you really get stuck NOAO offers a limited amount of technical support. Make sure you have done your homework and ask intelligent questions. Do *not* call and ask how to install the program, etc.

Each package and program has a set-up file that can be opened with **epar** <package/program name>. The set-up file opened with **epar** usually consists of a list of variables to be set by the user. For some variables there are limited choices and typing the first few letters will enter that choice. See the help files for an explanation of all the variable choices. Even though many find IRAF to be intimidating, the best part is you only have one place to look to fix problems. **epar** usually allows you to fix whatever is wrong and is your first step in trying to get things to work right. One up-front fix is to type **epar kpnocoude** and set Dispaxi = 1 and ^Z to update and save. This will look for spectra that run horizontally in your FITS images.

Data Reduction

The following steps outline the data reduction pipeline. In reality 50 images can be put through the pipeline to finished pixel shifts in about fifteen minutes. IRAF also prompts the user along the way for what to enter next, so the new user should not be intimidated by the program.

You should be starting with a series of raw exposures that we will progress through the data reduction steps. First do more than three and not more than about 20 dark frames at the same exposure time and temperature. These frames will be median combined in the MaximDL calibration routine and subtracted later from the raw images. The noise in the dark frames will be reduced by the square root of the number of exposures. IRAF also allows this type of processing and it too will produce a master dark frame but Maxim is easier.

Flat frames are taken with a full spectrum white light source down all the fibers. The exposure time is scaled to approximately half saturate the CCD pixel counts. Again do multiple flats and enter into MaximDL. The **mkflat** program in IRAF will do a more sophisticated flat but must be played with to get right. Finally run a full calibration on the raw image set which will subtract the master dark and flat frames.

The calibrated image set is ready to be manipulated in IRAF. The files can be transferred into the UNIX computer via FTP. IRAF can process FITS files directly but it is somewhat more convenient to convert them to IRAF specific imh files. In this way if something goes wrong you just delete the whole set and reconvert the FITS images. To convert FITS images us **rfits *.fit** and for a file extension type = a0. This will convert all the files to sequentially numbered images starting with a0001.imh. Most IRAF commands will take wild cards and do batch processing. imh files are actually two separate files in different directories linked through the program. Always use **imdel** when deleting imh files in order to keep a clean house.

You are now ready to define your "apertures". Each aperture defines the area on the image that one spectrum covers. This is how the program separates out the different stellar and reference spectra. **apedit** is used to create the defined areas. One of the flat frames should be used as the definition file by typing **apedit** <flat.imh> and say "yes" to editing apertures. This will produce a plot image with a peak for every spectrum stripe. The program will automatically try and define the apertures around the taller peaks. If they are not right move the cursor to the bracket that defines the width and type "**d**" to delete the aperture. Leave the cursor on the peak and press "**m**" to define a new aperture. "z" will automatically fit the aperture width to the peak. If necessary adjust the upper and lower limits with the cursor and the "u" or "l" keys. "q" and say "yes" to writing apertures to finish the programming.

Spectrum extraction comes next using the program **do3fibers**. This program allows for multiple spectra to be extracted from the same image. First **epar do3fiber** and fill out the read noise, photons per pixel count and flat file used to define the apertures. Typing **do3fiber** <file.imh> will bring up more questions. Say "yes" to resizing and editing the apertures and it will put you into **apedit** automatically. Redefine the apertures as described previously and press "q" to begin extracting spectra. The program will extract the spectra to a new file with the same name but .ms.imh file extension. This identifies them as multispectrum image files. View the multispectrum files with **splot** <file> and enter which aperture to plot. This program creates a line plot of the spectrum in a new window. The keystroke "t" will fit continuum and flatten the baseline, "(" will bring up the next spectrum in the image and "?" will bring up help in the XGterm window. **splot** is convenient to use with wild cards so a group of files can be viewed one right after another by pressing "q".

The final step is Fourier cross-correlation which is handled by the **fxcor** program found in the **rv** package. The closer the two spectra are in velocity, the better the program will resolve the spectrum shift. Your template spectrum is the base spectrum that the rest will be correlated against. Start with **epar fxcor** and set template = <template filename>, continuum = both, rebin = smallest, pixcorr = yes, output = xx and interact = yes. Running the program is easy: type **fxcor** and enter the spectrum file name to measure, wild cards OK. Next the program will show the default template file you entered in **epar**; if OK <enter> to continue. A new window will appear with a single peak that represents the fit between the template and current spectra. The pixel shift is shown at the bottom. Pressing "q" will write the results to xx.txt which was specified in output and go on to the next spectrum. When finished the pixel shifts are in column form in the xx.txt file. The file can be imported to other programs such as Excel for convenience.

The pixel shift is calculated against the dispersion per pixel to get the angstrom shift and corresponding velocity. For single images with reference and stellar spectra use Excel to subtract the reference drift from the original reference template and subtract that from the stellar shift to compensate for small systematic shifts. For shifts below 1 km/s diurnal motion becomes a factor for exposures more than a couple of hours apart. The **rvcorrect** program in the **rv** package can be used to determine the proper offsets for all orbital motions. **hedit** and **observatory** are related programs used to modify headers and enter time and location, etc.

Sources of Error

A good test of stability is to run continuous exposures with the reference lamp down all the fibers and cross-correlate the whole batch. When plotted this will show any systematic drift of the system. For testing system sensitivity a good challenge is to look for diurnal motion by shooting the same star through the evening. Accurate guiding is a must for precision measurements; if the star is offset on the fiber it can induce spurious shifts. In all, getting to know your system is the best way to achieve superior results.

Summary

With a good spectrograph and the power of IRAF software, the motion of the universe is open to the amateur. Spectroscopic binaries, comets, quasars, the Sun and extrasolar planets all make interesting and challenging targets. The advent of the CCD has brought true astrophysics within reach of many amateurs. For more information please visit www.spectrashift.com.

Helpful Command Glossary

hedit	Used to edit image headers
emac	Unix text editor
slist	List spectrum header
e	Allows arrows to retrieve the last command line, with up and down arrows
unlearn	Resets package to default settings

man -k \<keyword\>	Searches for text string in the help files
specplot	Plots multiple spectra together
bye	Leaves program package
pwd	Shows current path
package	Lists current loaded packages
help \<program name\>	Lists help file for that program.

Suppliers

[1]   Brass insert supplier:
McMaster Carr
multiple locations in USA
www.mcmaster.com
Knurled Press Inserts 1/2" long
part# 92395A116

[2]   Optics supplier:
Edmund Scientific
101 E. Gloucester Pike
Barrington, NJ 08007-1380 USA
899-363-1992
www.edmundoptics.com

[3]   Diffraction grating manufacturer:
Diffraction Products
9416 W. Bull Valley Rd.
PO Box 1030
Woodstock, IL 60098 USA
(815)338 6768

[4]   Chiller manufacturer:
OASIS Corporation Headquarters
265 North Hamilton Road, PO Box 13150
Columbus, Ohio, USA 43213 0150
614-861-1350 800-646-2747
www.OasisWaterCoolers.com
model RLF12-100

[5]   Temperature controller manufacturer:
Watlow Control
1241 Bundy Blvd
PO Box 5580
Winona, MN 55987 USA
507-454-5300
www.watlow.com
Series 965

[6]   Thorium argon lamp supplier:
Caleco Scientific
175 West Wieuca Rd Unit 139

Atlanta, GA 30342 USA
404-255-6672
www.caleco.com
part # p858A Thorium / Argon lamp
[7]   Thor/Ar power supply manufacturer:
Photron
3 Vesper Dr Unit 5
Narre Warren, 3805 Australia
Tel: 61 3 97049944
www.photron.com.au
model P209
[8]   Fiber optics supplier:
Polymicro
18019 N. 25th Ave.
Phoenix, AZ 85023 USA
602 375 4100
www.polymicro.com
[9]   Brake cable supplier:
Lexco
2738 W. Belmont Ave
Chicago, IL 60618 USA
800 626 6556
www.webservintl.com/testsites/thomas/Lexco cd/
home.htm
1/4 inch OD, armored, Teflon lined, low flex cable
housing.
[10] Hypo needle supplier:
The Micro Group
7 Industrial Park Road
Medway, MA 02053 USA
800 255 8823 or 508 533 4925
www.microgroup.com
20 gauge tube, 304 alloy, .0355" OD, .0025" wall
thickness, .0305 ID.
[11] WWW.spectrashfit.com – home of the author's
website on radial velocity and detection of
extrasolar planets.
[12] Cyanogen Productions
25 Conover Street
Ottawa, ON K2G 4C3
Canada
E mail: cyanogen@cyanogen.com
www.cyanogen.com
[13] NOAO National Optical Astronomy Observatory
950 North Cherry Avenue
PO Box 26732
Tucson, Arizona 85726
Phone: (520) 318 8000, Fax: (520)318 8360

www.noao.edu

IRAF homepage http://iraf.noao.edu/iraf/web/iraf homepage.html

Spectroscopy documentation http://iraf.noao.edu/docs/spectra.html

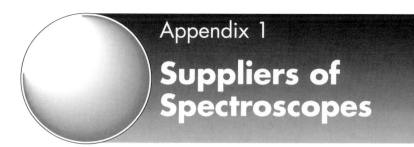

Suppliers of Spectroscopes

CVI Spectral Instruments
111 Highland Drive
Putnam
CT 06260
Tel: +1 860 928 1928
Fax: +1 860 928 1515
Email: cvilaser@neca.com
Web: http://www.cvispectral.com/

Ocean Optics Incorporated
380 Main Street
Dunedin
FL 34698
USA
Tel: +1 727 733 2447
Fax: +1 727 733 3962
Email: Info@OceanOptics.com
Web: http://www.oceanoptics.com/homepage.asp

Ocean Optics Incorporated (European Office)
Nieuwgraaf 108 G
6921 RK DUIVEN
The Netherlands
Tel: +31 26 319 05 00
Fax: +31 26 319 05 05

Optomechanics Research Inc.
P.O. Box 87
Vail
AZ 85641
USA
Tel: +1 520 647 3332
Fax: +1 520 647 3312)
Web: http://www.optomechanicsresearch.com

Rainbow Optics
1593 "E" Street
Hayward
CA 94541
USA
Tel: +1 510 581 8266

Santa Barbara Instruments Group
147-A Castilian Drive
Goleta
CA 93117
USA
Tel: +1 805 571 7244
Fax: +1 805 571 1147
Email: sbig@sbig.com
Web: http://www.sbig.com

Sivo Scientific Company
1400 Manhattan Avenue
Union City
NJ 07087
USA
Tel: +1 201 583 0279
Email: jmsivo@sivo.com
Web: http://www.sivo.com/

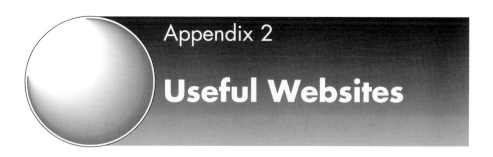

Appendix 2

Useful Websites

Amateur Astro-Spectroscopy
http://users.erols.com/njastro/barry/bar-page/aspectro.htm

CAOS Home Page
http://www.eso.org/~gavila/

Catalogue of Bright Star Spectra
http://users.erols.com/njastro/faas/pages/starcat.htm

CCD spectroscopy (Dale Mais)
http://members.cts.com/cafe/m/mais/

Christian Buil's Home Page
http://www.astrosurf.com/buil/

COAA
http://www.ip.pt/coaa/software.htm

Forum for Amateur Astro-Spectroscopy
http://users.erols.com/njastro/faas/

History of Astronomical Spectroscopy
http://www.achilles.net/~jtalbot/spectra/

MaxIm, Cyanogen Productions Inc
http://www.cyanogen.com/
Mini Spectroscopes

http://www.uwm.edu/~aawschwab/specweb.htm.
Sivo Scientific Company
http://www.sivo.com/SivoSci/sscindex.htm

Steve Dearden's Spectroscopy
http://www.astrosurf.com/dearden

Santa Barbara Instrument Group
http://www.sbig.com

Spectrohelioscope
http://home.ust.hk/~westland/spectroh.htm

Spectroscopy (Maurice Gavin)
http://www.astroman.fsnet.co.uk/spectro.htm

Spectroscopy with Small Telescopes
http://herry.me.uiuc.edu/astro/astro.html

The Fredrick N. Veio Spectrohelioscope Image Archive.
http://sunmil1.uml.edu/eyes/veio/index.html

Visual Spec (Valerie Desnoux)
http://valerie.desnoux.free.fr/vspec/

Appendix 3
Selected Bibliography

General

Karttunen, H. et al. (1996) *Fundamental Astronomy*, 3rd edn. Springer-Verlag, ISBN 3540609369.

North, G. (1980) *Advanced Amateur Astronomy*, 2nd edn. Cambridge University Press, ISBN 0521574307.

Payne-Gaposchkin, C. (1956) *Introduction to Astronomy*. University Paperbacks.

Van Zyl, J. (1996) *Unveiling the Universe*. Springer-Verlag, ISBN 3540760237.

Zelik, M. and Gregory, S. (1998) *Introductory Astronomy and Astrophysics*, 4th edn. Saunders College Publishing, ISBN 0030062284.

Historical Aspects

Hoskin, M. (1999) *The Cambridge Concise History of Astronomy*. Cambridge University Press, ISBN 0521576008.

Sullivan, N. (1965) *Pioneer Astronomers*. Scholastic Book Services.

Spectroscope Optics

Duncan, T. (1982) *Physics*. John Murray, ISBN 0719538890.

Hecht, E. (1972) *Optics*, 3rd edn. Addison-Wesley, ISBN 0201838877.

Nelkon, M. and Parker, P. (1994) *Advanced Level Physics*, 7th edn. Heinemann Educational Books, ISBN 043592303X.

Pedrotti, F. and Pedrotti, T. (1993) *Introduction to Optics*, 2nd edn. Prentice-Hall, ISBN 0130169730.

Sidgwick, J.B. (1971) *Amateur Astronomer's Handbook*, 3rd edn. Dover, ISBN 0486240347.

Stellar Spectroscopy

Cox, J.P. and Monkhouse, R. (1995) *Philips Colour Star Atlas.* Philips, ISBN 0540063169.

Kaler, J. (1997) *Stars and Their Spectra.* Cambridge University Press, ISBN 0521585708.

Kitchin, C.R. (1995) *Optical Astronomical Spectroscopy.* Institute of Physics Publishing, ISBN 0750303468.

Jaschek, C. and Jaschek, M. (1987) *The Classification of Stars.* Cambridge University Press, ISBN 0521389968.

The Contributors

Dr Steve Dearden is the laboratory manager with Saint-Gobain Vetrotex International, a manufacturer of glass fiber reinforcement products for polymers, situated in Chambéry, France. He has a Batchelor of Science degree in chemistry from Sheffield University, UK, and a PhD in physical chemistry from Bristol University, UK. A lifelong amateur astronomer, for the last few years he has been actively involved in developing spectroscopic techniques for use in amateur astronomy. He is married to Marie-Thérèse, who is an immunologist, and they have two teenage sons, Anthony and Christopher. Having recently moved to France from the USA, Steve now faces the task of setting up his "spectroscopic" observatory all over again! They live on the outskirts of the city of Lyon in southeast France. He can be reached at Steve.Dearden@wanadoo.fr or steve.dearden@saint-gobain.com His website is at: http://www.astrosurf.com/dearden

Dr Nick Glumac is an associate professor of Mechanical Engineering at the University of Illinois in Urbana-Champaign. His research areas include combustion and laser spectroscopy of reacting flows. Dr. Glumac has been active in amateur astronomy since 1997 when he and Dr. Joseph Sivo took a spectrum of comet Hale–Bopp that appeared in *Sky and Telescope* magazine. Later he and Dr Sivo developed a simple fiber-optic spectrometer and used it to demonstrate some of the possibilities for the amateur astronomer to perform spectroscopic measurements with a small telescope and CCD camera. This work was summarized in the February 1999 issue of *Sky and Telescope*. The original fiber-optic spectrometer design was modified and commercialized by Sivo Scientific Co., and this modified design was selected as one of the "Hot Products for 1998" by *Sky and Telescope*. Dr Glumac has subsequently designed and built a number of spectrographs for amateur astronomy use. Two of his designs are currently being used by Prof. Womack at St Cloud State University for comet research. An overview of his spectrographs and the spectra taken with them can be found at http://128.174.125.24/astro/astro.html, and he can be reached by e-mail at glumac@uiuc.edu.

Tom Kaye is 44 years old and during sunlight owns a manufacturing company in Chicago, Illinois. He is primarily interested in the scientific and research areas of astronomy and is one of the few amateur members of the American Astronomical Society. He is currently doing research on the possible connection between gamma ray bursts and extinctions in the fossil record. Spectrashift.com is the internet home of Tom's team of amateur extrasolar planet hunters. They were the first amateurs to resolve the orbit of an extrasolar planet using radial velocity methods similar to professionals. Tom is leading this team in building a 1.1 meter telescope and spectrograph with the intent to discover extrasolar planets using high precision radial velocities. The statement "science is for everyone" is Tom's phrase to live by. His website is at: http://www.spectrashift.com/

Dale E. Mais has been involved in amateur astronomy most of his life. He is an endocrinology researcher working for a biotech company in the San Diego area. Whilst his biology and chemistry degrees serve him well in his professional life, it is his chemistry background that he is enjoying as applied to spectroscopy. He is fortunate to have an observatory with a Celestron 14 as his primary instrument, CCD cameras and an AstroPhysics 5.1 inch (which he waited patiently for two years to obtain). His location 12 miles from Mount Palomar means that he benefits from the outstanding seeing, and relatively dark skies which the Hale telescope benefits from. His primary interest is spectroscopy and its application toward understanding the composition and other physical parameters of astronomical objects. In particular, he is doing a spectroscopic survey of C and S type stars, which often have abnormal heavy metal and/or isotope composition compared to solar system values. In addition, he is interested in quantitation of atomic/ionic species in stellar atmospheres. He can be reached at dmais@ligand.com His website is at: http://members.cts.com/cafe/m/mais/

Jack Martin has been an amateur astronomer for 30 years, with a special interest in spectroscopy, which started in the early 1990s. He began using a 35 mm SLR camera and direct vision spectroscope to take the spectrum of the Sun and terrestrial light sources. He then purchased a Rainbow Optics star spectroscope and began observing the spectra of stars. Then he made some simple modifications to his 0.30 m Dobsonian telescope, which enabled him to take photos of stellar spectra. This led to writing articles for the *Journal of the British Astronomical Association* and *Astronomy Now* as well as sending pictures of spectra for publication in the photo gallery. He has also developed darkroom-processing techniques for spectrograms and black and white slides of the same, and undertakes project work. Every year he does a display of his work at the European Astrofest, which attracts interest from other amateurs. He thanks Dr Mike Dworetsky, Director of the University of London Observatory, for his help, advice and encouragement over the years. He can be contacted at: jackmartinuk@hotmail.com

David Randell is by profession a computer scientist and Research Fellow at the Department of Electronic and Electrical Engineering, Imperial College, London, United Kingdom. He has been a keen amateur astronomer since childhood, and maintains a continuing interest in visual astronomy, optics and instrumentation. He is also an active member and webmaster of the Hanwell Community Observatory, based at Hanwell, Oxfordshire, United Kingdom (www.hanwellobservatory.org.uk), a project specifically set up to promote the general awareness of and public education in astronomy. His interest in spectroscopy arose literally by chance. When considering how best to use a CD as a grating in a DIY spectroscope, he noticed the tell-tale sign of absorption lines crossing the spectral streak formed by reflected sunlight falling on the compact disk. It was this single chance observation that eventually led to the chapter that appears in this volume. He can be contacted at: david.randell@hanwellobservatory.org.uk

Stephen Tonkin is a teacher of astronomy, physics, mathematics and drama. His interest in astronomy dates from his childhood, which was spent under the dark skies of tropical Africa. He makes most of his own astronomical kit and delights in showing others how effective kit can be made from what most people consider to be junk. His interest in stellar spectroscopy was triggered on a wet January evening in London in 1984, when he noticed, through his umbrella, the diffraction patterns from the light of street lamps. The resulting train of thought led to his first "junk" spectroscope, which used an old vinyl record as a grating; he has since used CDs and discarded binocular bits, and more recently acquired a Rainbow Optics grating. He has edited or authored several books and articles on astronomy and is a past chairman of the Wessex Astronomical Society. He can be contacted at: sft@astunit.com His website is at: http://www.astunit.com/

Index